W9-CBH-458

Shelton State Libraries
Shelton State Community College

DISCARDED

HF
5415,1265
W45
2009

THE NEW COMMUNITY RULES:
MARKETING ON THE SOCIAL WEB

Tamar Weinberg

DISCARDED

O'REILLY®

Beijing · Cambridge · Farnham · Köln · Sebastopol · Taipei · Tokyo

THE NEW COMMUNITY RULES: MARKETING ON THE SOCIAL WEB

by Tamar Weinberg

Copyright © 2009 Tamar Weinberg. All rights reserved.
Printed in the United States of America.

Published by O'Reilly Media, Inc., 1005 Gravenstein Highway North, Sebastopol, CA 95472.

O'Reilly books may be purchased for educational, business, or sales promotional use. Online editions are also available for most titles (*http://my.safaribooksonline.com*). For more information, contact our corporate/institutional sales department: (800) 998-9938 or *corporate@oreilly.com*.

Editor: Colleen Wheeler
Production Editor: Loranah Dimant
Copyeditor: Amy Thomson
Proofreader: Rachel Monaghan

Indexer: Ellen Troutman Zaig
Cover Designer: Karen Montgomery
Interior Designer: David Futato
Illustrator: Robert Romano

Printing History:

June 2009: First Edition.

Nutshell Handbook, the Nutshell Handbook logo, and the O'Reilly logo are registered trademarks of O'Reilly Media, Inc. The New Community Rules, the cover image, and related trade dress are trademarks of O'Reilly Media, Inc.

Many of the designations used by manufacturers and sellers to distinguish their products are claimed as trademarks. Where those designations appear in this book, and O'Reilly Media, Inc. was aware of a trademark claim, the designations have been printed in caps or initial caps.

While every precaution has been taken in the preparation of this book, the publisher and author assume no responsibility for errors or omissions, or for damages resulting from the use of the information contained herein.

ISBN: 978-0-596-15681-7
[Vicks Litho]   [2/10]
1245784904

To my husband, Brian, who still seemed to
tolerate me during crunch time.
I love you.

To our son, David Jacob, who arrived six weeks
early and was by my side (or inside) the most
as I was writing this book.
Welcome to the world, Little Man!

CONTENTS

FOREWORD

Dave McClure

Your lights are on, but you're not home
Your mind is not your own
Your heart sweats, your body shakes
Another kiss is what it takes
You can't sleep, you can't eat
There's no doubt, you're in deep
Your throat is tight, you can't breathe
Another kiss is all you need

Whoa, you like to think that you're immune to the stuff (oh yeah)
It's closer to the truth to say you can't get enough
You know you're gonna have to face it
You're Addicted to Love

—*Robert Palmer, "Addicted to Love" (1985)*

Hello…my name is Dave, and I'm a Facebook-aholic ("hi Dave, keep coming back!"). The rest of you may not be addicted to social media the same way I am, but I guarantee you it's only a matter of time. Now that Oprah and CNN have become run-of-the-mill street-corner pushers for social media crackpipes like Facebook and Twitter, you can bet the rest of the consumer mainstream ain't far behind. Mark my words, folks: we're all being seduced by a dangerous and sexy online mistress named Social. If you haven't fallen for her yet, you will.

If you have ever read a blog, visited a MySpace page, watched a YouTube video, checked out a photo on Flickr, or clicked on a link in Twitter, then five hours later, looked up to check the clock and realized it was 4:00 AM, you know what I mean. Admit it, you've been there: heaven help me, the baby is screaming and needs a diaper change, but gimme a sec, I just need to click on one…more…link…aaah. Now doesn't that feel better?

You might be a teenager on Hi5, profile-hopping all the hot girls in your freshman class at high school, or a grandmother anxiously checking YouTube to see if your daughter has uploaded the latest video of your three-year-old grandson. You might be a punk rocker adding a new song to your band's MySpace page or a Harvard grad surfing LinkedIn to see who you know at Google who's hiring. You might be the Real Shaq Daddy tweeting out nightly box scores and a slam-dunk on Yao Ming, or Barack Obama rallying the faithful to get out to vote via SMS on the eve of the most historic election in American history. From the largest to the smallest, from the youngest to the oldest, the world has become engrossed, enthralled, and addicted to social media.

Unless you've been in a coma for the last five years, your behaviors and interactions with social media have changed dramatically. We now spend more time connected—both literally and figuratively—than ever before. Our offline-online existence is fused together into an electronically enhanced experience that would have seemed unbelievable just over a decade ago, but now seems almost second nature. One wonders how people ever managed to make plans to meet up for dinner or a night out on the town before everyone had email, eVite, Yelp, or text messaging. Our fascination and fastened-nation with all things digital has been both a blessing and a curse, allowing people to communicate whenever and wherever they please, even if that means listening to the sales guy in the bathroom stall next to you talking to a customer and wondering if you should wait 'til he's done before you flush.

The first 10 years of the Internet Revolution were all about getting computers connected to the World Wide Web. But the next 10 years are going to be all about getting people connected to one other. There are now over 1 billion people online across the globe, and over 3 billion people with mobile phones who can send a text message. Imagine how much time we can all waste poking one other on Facebook!

More seriously, this sea change in how people spend their lives and leisure hours has created a challenge for those in traditional marketing roles. As with the explosion of cable television channels in the 1990s and subsequent fragmentation of mass market media and advertising, online behavior in the 21st century has been moving away from large portal mass-produced websites like AOL and Yahoo!, and toward a world filled with search engines, social networks, millions of tiny blogs and "long-tail" websites, user-generated content sites, news feeds, apps, widgets, RSS, email, SMS, IM, chat, Twitter, bookmarks, etc, etc. Finding ways to effectively reach customers in the world of Web 2.0 has become a Sisyphean task, requiring a wide variety of online marketing skills and an endless number of communication channels.

And yet there also exists the everyday miracle of one clever, creative individual who executes a very cheap, viral, word-of-mouth campaign that reaches millions overnight. How can this be? We are both powerless and powerful at the same time. We are fragmented and yet unified. We are solitary shut-ins glued to our computers, but we are powerfully and instantly connected to thousands of others all over the Earth. We are billions of people on the World Wide Web, and we are a billion people blathering on in a billion and one tongues.

This is social media. And like the social beings who create it, social media is messy and confused. It was in the middle of that mess that my personal journey began. Let me explain….

Back in late summer 2001, I had the good fortune of accepting a job offer at PayPal, while the rest of the dot-com world was crashing all around me. Little did I know the towers of the World Trade Center in Manhattan would also come crashing down my first day on the job. While still in shock at a changed world offline, I began putting my toe in the waters of a brave new world online as well.

I had always been a geek of some kind—music geek in grade school, math major in college, computer programmer after graduating, and a small-time Internet entrepreneur in the mid-1990s until my company got acquired in 1998. However, my new job at PayPal was in (developer) marketing—pretty unfamiliar territory for a geek. I wasn't even sure how I got the job; a friend who was a PayPal angel investor had referred me, since he knew I'd been organizing several Silicon Valley tech and entrepreneur user groups for many years. I guess PayPal figured that was as close as it could get to someone who knew how to market to developers, so it gave me a chance.

Now just to be clear, there is nothing more anathema to geeks and programmers than someone who has a business card with a marketing title, except perhaps someone in sales, but at least geeks understand salespeople are necessary to make money that pays their salaries. So basically, as far as developers are concerned, marketing folks are the absolute bottom of the food chain—they're assumed to be both clueless and useless, and liars to boot. As a former developer myself, I realized my job was going to be all about marketing our services to repressed loner, smart-ass geeks who thought I was a dumb, incompetent liar. Great.

Given the humbling and humiliating task ahead of me, and given my dirty little secret of not knowing one damn thing about traditional marketing, I realized I better come up with some pretty creative tricks/hacks…and fast. Hack #1: change the official job title on my business card from "Director of Marketing" to "Director of Geek Marketing" (disguise and subterfuge, become part of the community). Hack #2: stop trying to *sell* developers on PayPal, and just focus on helping them *use* the product and provide tech support, *listen* to what they were asking for, and see if I could get the product team to fix bugs and build something geeks would use. Hack #3: since they probably knew more than me, appeal to developers to help answer questions, and recruit geek advisors and promote them as experts to the rest of the community. Hack #4: get all of our technical documentation and code samples out in the open on a no-login-required site, without requiring anyone to create a PayPal account to learn. Hack #5:

start a message board and blog (had to bend some rules and avoid corporate bureaucracy, but I did it), and get an open channel of publishing and communication to the community.

I could go on, but I think you get the picture—let's just say I did some *very* nontraditional marketing in the first year or two. And I really had to change how I thought about marketing in order to reach the people I was going after. In fact, much of my success was due to subverting, bending, and even breaking the normal rules of corporate marketing to do what I needed to get done. And finally, I had to become part of the community itself, and I had to create some nontraditional publishing and communication models to engage the community to help me do my job.

Along the way in becoming a mole in the machine, I also discovered a number of other important new trends and techniques in online marketing: search marketing (both organic and paid), email newsletters and distribution lists, blogging, mini-apps and widgets, message boards and forums, RSS, screencasts, instructional video, social networks, and many, many other geeky pursuits that consisted mostly of me goofing off online and somehow getting away with saying I was doing real work. While it may have seemed like I was screwing around wasting a lot of time (cough, cough…nothing could be further from the truth!), it turns out I was getting some world-class on-the-job training in social media marketing. Who knew?

As I spent more time diving deep into this Ocean of Social, I realized something important was happening and changing how people were communicating. Starting somewhere between 2001 and 2005, a whole bunch of non-geeks were getting computers, getting digital cameras and mobile phones, getting broadband connections, and getting online. The Internet and the browser were just the beginning; by the time YouTube arrived in 2005, the Internet had already been taken over by the masses. By 2008, your mom or grandmother were probably stalking you on Facebook and trying to find out who you were hooking up with.

This was not your geeky old Internet—*this* was the glory of the World Wide Web, and people were doing a *whole lot more* of the following:

- Browsing the World Wide Web (from iPhones as well as computers)
- Using search engines (aka "The Google") to find all kinds of stuff
- Reading blogs, looking at pictures, listening to music, and watching videos
- Creating profiles and browsing and flirting and "poking" on social networks
- Sending messages and links via email, text/SMS, and Internet Messaging (IM) systems

As each of these activities in turn spawned entirely new ecosystems and communication channels dedicated to legions of fans, online populations similarly dedicated themselves to the creation and consumption of new media/social media in these online environments. Not only had we become addicted to the Network, we had *become* the Network:

We The People,

In Order to Form a More Perfect Platform,

Establish Internet Equality,

Ensure Domestic Social Connectedness,

Provide for the Creative Commons,

Promote the General Web-fare,

and Secure the Blessings of Liberty

to Our Blogs and Our Friends and Followers,

Do Ordain and Establish this Network

for the .COM, the .NET, the .ORG,

and the Entire World Wide Web!"

Well, maybe it didn't happen quite like that…but I bet you in a hundred years, people will look upon the creators of the Internet, search engines, social networks, and some of the more famous websites akin to the way older generations think about our founding fathers. I mean, didn't Al Gore invent the Internet? I rest my case.

And as we begin to explore what social media is about in the Second Age of Aquarius, I can think of no one more qualified to bring you kicking and screaming into the 21st century than Tamar Weinberg. Tamar is a friend, guru, and colleague who has been swimming in the ocean of search engines and social media for over 10 years, and her annual "Best Of" list of Internet marketing articles is a must-read for all things search, social, and beyond.

With no further ado: I bring you Tamar Weinberg, and the Social Web.

Social media marketing is more than just a buzzword. It's a way of life and a means of survival in today's Internet lifestyle. Whereas the Internet of the past was more about "me, myself, and I," the past few years have brought about substantial change: our online interactions are now more social. Our product purchases are often driven by user reviews. We enjoy reading interesting stories shared with us by our friends and colleagues. We have seen the rise of online communities where individuals with similar backgrounds or interests can connect to one another.

Regardless of whether you've done traditional marketing online or absolutely none at all, diving into this unfamiliar territory is not that much of a challenge. To understand the basics of "social media marketing," let's break down the terminology. The idea behind social media marketing is to leverage the "social" through its "media" (communication and tools) to "market" to your constituents.

The big idea behind social media marketing that you are focusing on is *communication*. Fortunately, communities exist that already have active participants—those passionate about a specific subject—and better yet, there are numerous tools that can help facilitate this kind of communication. If you're a small-business owner or even a member of a corporate entity but are unfamiliar with this territory, there are many ways to dive in and become part of the *conversation*.

Conversation is a two-way dialogue. Unlike traditional marketing, social media marketers are required to start listening and talking to their constituents. This is possibly the biggest hurdle facing a social media marketing initiative. However, have you searched for your product or brand name today? What are people saying? Don't you feel compelled to respond?

It's about time that you began understanding the social media landscape. It's about time that you began leveraging the networks where people are already conversing about you to respond favorably to their feedback or criticisms. It's about time that you embarked on a social media marketing initiative.

Organization of the Material

Chapter 1, *An Introduction to Social Media Marketing*, introduces the concept of social media marketing and explains its role in today's online marketing initiatives. This chapter also covers some of the primary tools used for a social media marketing campaign.

Chapter 2, *Goal Setting in a Social Environment*, discusses the challenges and hurdles faced in social media marketing and also explains the various ways to leverage social media marketing to achieve specific goals.

Chapter 3, *Achieving Social Media Mastery: Networking and Implementing Strategy*, outlines tools for monitoring online chatter and what you can do to appease your audience.

Chapter 4, *Participation Is Marketing: Getting into the Game*, explains how participation online is critical to success in social media marketing and presents case studies highlighting small and large businesses achieving success with this tactic. Chapter 4 also discusses another important part of social media marketing: reputation management.

Chapter 5, *Using Blogs to Communicate, Influence, and Learn from Your Constituents*, describes the growth of blogs and explains how to set up a blog from scratch and how to make it friendly for social media communities.

Chapter 6, *Microblogging Magic: How Twitter Can Transform Your Business*, illustrates microblogging service Twitter and explains how to use the service. It also features case studies of businesses that have successfully navigated the Twitter landscape for marketing gain.

Chapter 7, *Getting Social: Facebook, MySpace, LinkedIn, and Other Social Networks*, discusses the primary three social networking sites and explains how you can use them appropriately for social media marketing gain.

Chapter 8, *Informing Your Public: The Informational Social Networks*, highlights knowledge exchange websites such as Wikipedia and Yahoo! Answers and offers insights into how you can use these networks to establish thought leadership and expertise.

Chapter 9, *Leaving Your Mark: How to Rock the Social Bookmarking Space*, presents the concept of social bookmarking sites and explains how to use these services.

Chapter 10, *Social News Brings You Page Views*, explains the benefits of content creation for social news sites and outlines the steps you can take to write great content for, and to become a successful contributor, to those sites.

Chapter 11, *New Media Tactics: Photography, Video, and Podcasting*, covers the services that allow you to promote your photographs and videos and explains how you can become a rockstar podcaster or videoblogger.

Chapter 12, *Sealing the Deal: Putting It All Together*, explains the best approach for a successful social media marketing strategy, especially once you're armed with the information presented in the preceding chapters.

Questions and Comments

Please address comments and questions concerning this book to the publisher:

O'Reilly Media, Inc.
1005 Gravenstein Highway North
Sebastopol, CA 95472
800-998-9938 (in the United States or Canada)
707-829-0515 (international or local)
707-829-0104 (fax)

We have a web page for this book, where we list errata, examples, and any additional information. You can access this page at:

http://www.oreilly.com/catalog/9780596156817

To comment or ask technical questions about this book, send email to:

bookquestions@oreilly.com

For more information about our books, conferences, Resource Centers, and the O'Reilly Network, see our website at:

http://www.oreilly.com

Safari® Books Online

 When you see a Safari® Books Online icon on the cover of your favorite technology book, that means the book is available online through the O'Reilly Network Safari Bookshelf.

Safari offers a solution that's better than e-books. It's a virtual library that lets you easily search thousands of top tech books, cut and paste code samples, download chapters, and find quick answers when you need the most accurate, current information. Try it for free at *http://my.safaribooksonline.com*.

Acknowledgments

When working on a book on social media, one must realize that the collective intelligence—the social—is of utmost importance for materializing this dream and bringing it to fruition. Without the help of several individuals, both for feedback and content, *The New Community Rules* would never have been possible.

With that said, there are a few individuals who must be thanked for their debate, their insights, and their ears while I spent the last few months writing what I hope to be an authoritative resource on social media strategy and the tools and communities to make that strategy a reality. In no particular order, I'd like to give thanks to these individuals: Jason Falls, blogger at Social Media Explorer (*www.socialmediaexplorer.com*), whose blog can supplement this reading as he has highlighted two case studies seen in this book; Jane Quigley, for insightful corporate strategy; Matthew Inman of 0at.org, who is an artist and a creative mind and is at the forefront of viral quiz and questionnaire technology; Andy Beal of Marketing Pilgrim (*www.marketingpilgrim.com*), for his expertise on reputation management and then some; Matt McGee of Small Business SEM (*www.smallbusinesssem.com*), for having great Flickr photos and providing wonderful insights into the content of this book; and Dave McClure, the same individual who wrote the foreword for this book, and one of the most amazing minds in this arena and whose background is not so far off from my own.

Special thanks to those who gave me some great insightful feedback during our interviews, both online and over the phone. These individuals include Tony Hsieh, CEO of Zappos.com; Nick Ayres, Interactive Marketing Manager of The Home Depot; Ed Nicholson, Director of Community and Public Relations at Tyson Foods; Rob Key, Constantin Basturea, and Paull Young of social media marketing and communications agency Converseon; Justin Levy, General Manager of Caminito Argentinean Steakhouse; Frank Eliason, Director of Digital Care at Comcast; Shashi Bellamkoda, Social Media Swami at Network Solutions; Morgan Johnston of JetBlue's Corporate Communications team; Michelle Greer of SimpleSpeak Marketing; Sam Feferkorn, consultant for Oh! Nuts in New York; Regan Fletcher, Vice President of Business Development at Yoono; and Andrew Milligan, owner of Sumo Lounge.

Additionally, many thanks to my eyes and ears: Anna Bourland, Brian Wallace, and Samir Balwani. Also, thank you to Loren Feldman, Jay Izso, Brent Csutoras, Chris Winfield, Allen Stern, Anita Campbell, Laura Fitton, Muhammad Saleem, Jonathan Fields, Todd Defren, Greg Davies, Joe Fowler III, and Brian Hill for your tidbits and advice.

Most of all, thanks to my husband, Brian, who most appropriately fits the dedication since he really had to endure all of these few months of hard work, but who did it with grace and was supportive throughout.

CHAPTER ONE

An Introduction to Social Media Marketing

Social media, which relates to the sharing of information, experiences, and perspectives throughout community-oriented websites, is becoming increasingly significant in our online world. Thanks to social media, the geographic walls that divide individuals are crumbling, and new online communities are emerging and growing. Some examples of social media include blogs, forums, message boards, picture- and video-sharing sites, user-generated sites, wikis, and podcasts. Each of these tools helps facilitate communication about ideas that users are passionate about, and connects like-minded individuals throughout the world.

According to the Universal McCann's Wave 3 report, released in mid-2008,[1] social media is rising and does not seem to be stopping anytime soon. Among all Internet users between the ages of 16 and 54 globally, the Wave 3 report suggests the following:

- 394 million users watch video clips online
- 346 million users read blogs
- 321 million users read personal blogs
- 307 million users visit friends' social network profile pages
- 303 million users share video clips
- 202 million users manage profiles on social networks
- 248 million users upload photos
- 216 million users download video podcasts
- 215 million users download audio podcasts

- 184 million users start their own blogs
- 183 million users upload video clips
- 160 million users subscribe to RSS feeds

Social media penetration seems to be a continuing trend.

Social media marketing (sometimes referred to by its acronym, SMM) connects service providers, companies, and corporations with a broad audience of influencers and consumers. Using social media marketing, companies can gain traffic, followers, and brand awareness—and that's just the tip of the iceberg.

THE INTERNET EVOLUTION AND HOW IT RELATES TO SOCIAL MEDIA MARKETING

Two decades have passed since Tim Berners-Lee invented the World Wide Web. Initially designed for the physics community,[2] Berners-Lee likely never imagined that his project would later become known as the "information superhighway" and that the Internet would end up interconnecting millions of computers worldwide, providing vast amounts of information to individuals. Berners-Lee likely never imagined that the Internet would be accessible to every household and that it would facilitate communications throughout the world. In the last few years, the Internet has evolved into a "social web," connecting like-minded individuals with communities that allow them to express themselves and engage in lengthy debates at any time of the day.

Ask.com, Lycos, Metacrawler, Altavista, Google, Microsoft Live, Yahoo!, and other search engines were created with the intention to organize the world's information. A new discipline known as *search engine optimization* (also referred to by its acronym, SEO) became mainstream among marketers who wanted to understand the nuances of how a search engine wound rank results for various search phrases. The goal of a search engine optimizer was to have the pages of his client's website appear on the first page of search engine results. For example, if a client specialized in the sale of "blue fish" and an individual was using a search engine to find a "blue fish," a search engine optimizer would want his client's site to show up first in the results.

Individuals are forever searching for information, and search engine optimizers help organize content on a web page so that their clients' websites rank higher than the competition's. Search engine optimization typically involves the analysis of elements on a particular web page and enhances them, using available search engine algorithmic knowledge (as seen by repeated success and observation; search engines naturally keep their algorithms top secret) for heightened visibility in the search engine results.

Search engine optimization is part of a larger picture, *search engine marketing*, which encompasses a variety of other tactics for heightened awareness in the search engines. Before social media marketing made its foray into the marketing arena, search engine marketing integrated these major components:

- *Search engine optimization*, which focused on on-page factors, including title tags, metatags, keyword research, and other techniques.
- *Link building*, an offsite promotional tactic to build quality links from other websites to improve rankings.
- *Pay-per-click*, a model that allowed individuals to bid on clicks and to pay for high rankings. In this model, search engine users saw "sponsored" listings alongside regular "organic" results. It was typically much easier for businesses to achieve high rankings in this area: the more money invested in the campaign, the more visibility to the casual surfer (contingent upon other algorithmic factors).

DEFINITION
Most search engines contain listings that consist of paid advertisements (sponsored listings) and unpaid listings, where the placement is based on a highly classified search engine algorithm that often relates to relevancy, number of inbound links, and other data points. *Organic listings* are these unpaid results that often show up on the left side of the search engine results page.

Where We Are Now

Until recently, the Internet was largely an informational medium. However, in the last couple of years, the Internet has become increasingly social. We are now looking at websites, habits, and behaviors of our peers in order to make well-informed and educated decisions about our next move, be it a buying decision or another endorsed article to read late at night. Websites such as MySpace and Facebook have emerged to make communication between peers fast and easy. That's only the tip of the iceberg, though. Social websites have been built to unify individuals with similar interests: social news sites that are governed by the "wisdom of crowds," social bookmarking sites that allow individuals to discover websites that a large number of people have already discovered, and niche social networks that unify individuals under a common interest. As such, a new discipline, *social media optimization*, also called *social media marketing*, has evolved.

What Is Social Media Marketing?

Social media marketing is a process that empowers individuals to promote their websites, products, or services through online social channels and to communicate with and tap into a much larger community that may not have been available via traditional advertising channels. Social media, most importantly, emphasizes the collective rather than the individual. Communities exist in different shapes and sizes throughout the Internet, and people are talking among themselves. It's the job of social media marketers to leverage these communities *properly* in order to effectively communicate with the community participants about relevant product and service offerings. Social media marketing also involves listening to the communities and establishing relationships with them as a representative of your company. As we will discuss later in this book, this is not always the easiest feat.

> **DEFINITION**
>
> The term *social media optimization*, which many today equate with *social media marketing*, was coined in 2006 by Rohit Bhargava.[3] Bhargava explained the concept of social media marketing as optimizing a site in such a way that written content garners links, which essentially acts as a trust mechanism and endorsement. Social media optimization also helps build brand awareness and raise visibility for the marketed product or service.

In essence, social media marketing is about listening to the community and responding in kind, but for many social media marketers, it also refers to reviewing content or finding a particularly useful piece of content and promoting it within the vast social sphere of the Internet.

Social media marketing is a newer component of search engine marketing, but it is really in a class of its own. It does not relate only to searching; it relates to a broad class of word-of-mouth marketing that has taken the Internet by its horns. Fortunately, the phenomenon is only growing at this point.

In the end, social media marketing can achieve one or many of the goals listed in the following sections.

Bringing Traffic to Your Website

Using available social media tools, users endorse approved content for their peers. As soon as an active user of a social news site or influencer discovers a piece of content and spreads it, word of mouth commences. The idea is a viral spread, which is heightened by online communities and the cross-pollination of content on other social media sites. Figure 1-1 illustrates this phenomenon.

Driving Relevant Links to Your Website

Considering that link building is a big part of search engine marketing, social media marketing eliminates the need to seek out a costly link-building expert and can help build organic links. When a blogger or website owner discovers a relevant piece of content, the natural instinct is often to share the content on the website or blog with a direct link to the piece of discovered content. These links in turn help to communicate to search engines that the blogger or webmaster has made a decision to endorse the web page, as its content is considered trustworthy. As many search engine marketers can attest to, the more links to your site, the

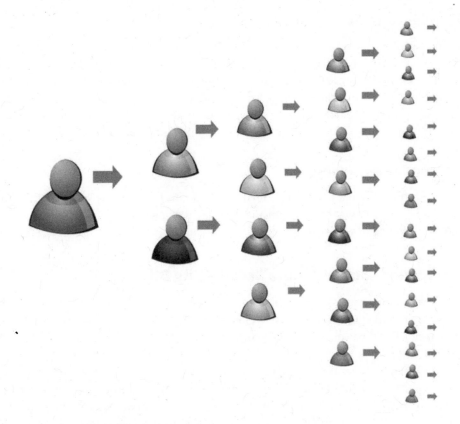

FIGURE 1-1. A graphical representation of viral marketing

more opportunities you have to be discovered by both readers and visitors, as well as users looking for related content through searches performed on search engines. Links enhance discoverability. Social media sites are just a starting point, but with the right content, the gift of compelling social media content has the potential to give back to the content creator twenty-fold or more.

Making Consumers Brand-Aware

Obviously, a strong market presence is beneficial for getting business from customers who need your product or service today. However, creating brand awareness today can also help you in the future. Consumers who become aware of your brand now, even if they aren't actively seeking your product or service, are likely to remember you in the future and seek you out when they actually do need your product or service. If you leave a positive first impression on your diverse audience, you will likely reap benefits from exposing it to your product early, especially since one of the key ideas of social media marketing is recommendations: the idea behind social media is that friends recommend links, websites, and products to their peers.

Driving Conversions

Given a compelling marketing strategy and creative demonstration, social media marketing can lead people to purchase the desired product or service. On the contrary, a poor marketing demonstration will likely cause the consumer to distance himself from the advertised product. Consider this logic: if you are selling a software product offering and decide to innovate with a poor-quality video laden with mistakes and monotonous voiceover, how likely is that video to contribute to increased sales? Presentation and layout are crucial in social media marketing.

Triggering Conversations

If you're getting links out of your social media strategy, it's because people are talking about you. It's important to note that social media users are seeking out the voices of their peers—they're not listening to you as a corporate entity anymore.

What Makes "Social Media" Marketing Different?

Now that we've established some of the benefits of social media marketing, it should be clear that traditional marketing tactics are not as effective as they once were, because consumer trust in these media forms has declined. Today, information is more easily accessible online, and more significantly, that information is a lot easier to find. Generations are becoming increasingly digital-savvy. Text messaging and web activity are becoming second nature (and are claiming addicts on a daily basis). If a consumer is seeking out information about a particular product, she won't necessarily sit down with a cup of coffee and read her favorite magazine to find information about the product; she is more likely to turn on her computer and look for reviews and endorsements from other individuals just like her.

Social media marketing is a promising evolving technology with much potential, and there are successful case studies to back up that sentiment, many of which we will explore in this book. However, there are other reasons to engage in a solid social media strategy in addition to (or instead of) traditional marketing strategies. These include:

Social media marketing facilitates natural discovery of new content
> Content crafted properly can be exposed to hundreds of new website visitors, from the casual surfer to the extreme enthusiast, in a very spontaneous fashion. Unlike paid advertising, which is forced upon web surfers, social media lets visitors view content that is not necessarily associated with commercial intent. If I like a website because the marketing piece is hip, innovative, and genuine, I'll pass it on to my peers using social media sites and they'll pass it on to their peers because they also like it. Content can reach thousands of new eyeballs quickly without interfering with traditional marketing, but social media marketing does not interfere with other marketing strategies, either.

Social media marketing boosts traffic numbers

Traffic comes to websites from sources other than search engines, and many of those sources include social media sites. Once you have established yourself as a community participant worth following, people will be interested in what you have to share and will likely pass relevant your blog posts, videos, or articles onto their peers.

Social media marketing builds strong relationships

If you are genuinely paying attention to members of the communities that are part of your marketing message (or not even associated at all), you can build strong relationships when you take the time to respond to concerns or feedback. Even communities that are not necessarily related to your company, brand, product, or service offering have members who may individually be interested in knowing more about you and what you have to offer. And since it is so easy to spread your message via word of mouth online, if you really leave a good impression on those who you interact with on a regular basis, it's almost certain that they will recommend you to a peer who is seeking your service or product— that is, if they believe in it (and you!)

Social Media Marketing Is a Cheap Alternative to Traditional Marketing— with High Returns

You can certainly hire someone to provide strategy and tactics to bring you success, or you can integrate social media marketing in-house for much less (though before you do, read this book first!). Social media marketers still need to understand the rules of engagement, participate in communities on a regular basis, and capitalize on emerging trends. Such activity will prove to be highly cost-effective. Buying hundreds of links on untargeted sites, for example, may cost you thousands of dollars, but if you practice the creative strategy of social media marketing, the return on investment can be substantially higher. Plus, social media marketing has the added benefit of heightened awareness about product offerings.

Where is my return on investment?

Maybe you've decided to take the plunge and utilize social media marketing in-house. Maybe you've decided to hire a renowned social media marketing consultant to help implement and then execute your social media strategy. How are you going to measure your return on investment (ROI) to see if your investment was worth all the effort?

First, if you're trying to determine how much a social media marketing strategy should cost, there's no "one size fits all" approach. Social media marketing simply does not have a fixed cost. Depending on the scope of the project, social media can vary from hundreds of dollars to hundreds of thousands of dollars. When pricing out possible consulting engagements, never put all your eggs in one basket. Look for a fair mix of social sites and communication opportunities in the most ideal social media marketing campaign.

NOTE

You never want to work with a consultant who will give you visibility on Digg but ignore the other sites that exist for a similar purpose.

Now where is your ROI, and how can you tell if your social media marketing strategy is the right one? Social Media Explorer Jason Falls explains the issue of determining ROI for social media marketing:[4]

> The problem with trying to determine ROI for social media is you are trying to put numeric quantities around human interactions and conversations, which are not quantifiable.

Measuring success

Further, social media marketing results cannot be measured immediately. Your strategy does not work overnight, but rather works over the long term. Like any sort of marketing tactic, social media marketing puts your product or service in front of a group of users who will be interested in sharing the offering with their peers, though the process of sharing is only as rapid as the individuals who want to pass on the content. In an effective campaign, the results should speak for themselves.

In many instances, social media is also about listening and engagement. Over the long term, if you see more positive sentiment being expressed about your company, that should be a win in itself.

There is no "one size fits all" strategy that works with everyone. Each product and service is different. Each online community is different. By communicating with the right group of people online and then revising your strategy as needed based on feedback, you will likely see some incredibly valuable results that will help you sell your product—or you'll go back to the drawing board.

Later in this book, you will be introduced to tools that will help to measure success—all depending on your goals—and you will be able to tweak your campaigns effectively based on the response to your strategies.

Let's face it: the online world is becoming saturated. There are a substantial number of individuals now flocking to the Internet to find answers and get direction. It's time you talk to those people, especially since they may have questions about a product or service that you may have the answers for.

With social media marketing, you'll see that if your outreach efforts are received well, you'll gain a percentage of supporters. As these supporters spread the word about your offering, you will gain additional supporters. With the right targeting and proper message, gaining loyal followers will allow you to build up a group of individuals who will be willing and able to act when you launch a new desirable service offering. This is so much easier to do online because the message is so much easier to spread.

If you are looking for a sure-shot way to achieve fast results, this book is not for you. Like any marketing discipline, social media marketing takes diligence, effort, and persistence. By reading this book, you will understand how to:

- Establish goals for your social media marketing campaigns
- Create a strategy for executing your social marketing efforts
- Communicate effectively with the communities you intend to target
- Take charge of the conversation, even if it's not on your website
- Gain exposure from participating among many social channels
- Utilize social media to handle a reputation management crisis
- Utilize blogs and bloggers to send messages to larger groups of individuals
- Leverage existing sites to market your products
- Craft content that is currently "hot" within many social media circles

A Brief Introduction to Social Media Portals

Fortunately, there are already a number of portals available online that can empower you, as a marketer, to start spreading your message. Innovation, too, can bring success, though it's not the only way. There are already so many sites that have been built around the idea of the collective mindset, and it is your responsibility—if not your duty—to understand the communities that frequent these social sites and leverage them for your benefit while also giving back to the community.

> **NOTE**
> The emphasis here is that people want to know you are providing something valuable to them. Communities will not respond if your intentions are selfish in nature. Later in the book, we will discuss how to work with communities to spread the message.

With social media portals, your current and potential customers can associate themselves with you and your brand. They do this by bookmarking a page on a social bookmarking site, becoming your fan on a Facebook product page, and voting up a story on a social news site, among other tactics. In this section, you will learn some of the more popular social media portals—not at all intended to be an exhaustive list (as there are new ones cropping up from day to day). Later, you will learn how to leverage these networks to spread your message.

Social News Sites

Social news sites rely on the collective to vote on news stories that individuals think should be exposed to a larger audience. In essence, when a story is submitted to a site, it has one vote. The goal of social news sites is to get the story enough votes (which may vary per social news site) to hit the front page. Since thousands upon thousands of visitors often do not venture farther than the front page of social news sites, getting your story there can bring hundreds of thousands of visitors to your site in a short while, with the added benefit of getting targeted links from influencers. The reason for this is that popular social news sites are regularly visited by bloggers, journalists, and other influencers who try to find their writing inspiration from content that is on the front page of these sites. If the community already publicly endorsed this content, it's fair game for the writers to pass on to their readers. Some social news sites are covered in this section.

Digg (http://www.digg.com)

By far, Digg is the most popular site at the moment for sharing information socially. Digg was originally launched in late 2004 with an emphasis on technology news, but it changed its game plan in early 2008 to target a much wider audience.[5]

reddit (http://www.reddit.com)

Launched in 2005, reddit is known as the second most popular news site. reddit found big success in January 2008 when it launched subreddits, which enable users to create their own categories in which to submit stories.[6] As a reddit user, you can subscribe to specific categories and get the content that you want without the clutter of other news.

Mixx (http://www.mixx.com)

Mixx is an up-and-coming social news contender that was founded in late 2007. It is one of the smaller social news sites, but has an incredibly passionate and active community.

Social Bookmarking Sites

Social bookmarking sites allow you to store your favorite sites, often with metadata (tags, for example) to be retrieved at another time or in another place. While some people use social bookmarking just so that they can access their bookmarks from several computers without feeling tied down to any single location, social bookmarking also allows you to discover new content saved by your peers. By default, social bookmarks are public, though there are options to make the bookmarks private. The more popular social bookmarking sites are covered in this section.

delicious (http://delicious.com)

This social bookmarking giant is now owned by Yahoo!. The site boasts more than 5 million users and more than 150 million URLs. On July 31, 2008, delicious launched a newly redesigned site that boasted impressive speeds, enhanced sorting, and a stronger emphasis on networks.

StumbleUpon (http://www.stumbleupon.com)

StumbleUpon is a unique kind of social bookmarking site. It allows you to discover content using a toolbar. When you click Stumble!, you are shown a site tailored to your interests (per your specifications when you registered). You can then provide feedback to the service as to whether you like the content or not. Based on your feedback, StumbleUpon provides additional (or fewer) pages on the specific topic.

Social Networks

Social networks are the websites that you use to let individuals know exactly who you are or establish a profile to find others with similar interests. Often used to connect with old friends or to find new ones, social networks are some of the most popular sites on the Internet. Three of the key social networks are covered in this section.

Facebook (http://www.facebook.com)
College student Mark Zuckerberg launched Facebook in 2004 to allow other college students to keep in touch with their friends. Now Facebook is one of the most popular websites in the U.S. has been growing virally throughout the world.

MySpace (http://www.myspace.com)
MySpace is another popular social network. Founded in 2003, it has grown to over 100 million accounts and was acquired by News Corporation in 2006.

LinkedIn (http://www.linkedin.com)
Launched in mid-2003, LinkedIn is a network that connects professionals in all disciplines all over the world. LinkedIn is intended for those who are business-oriented, and is best described as a "virtual resume" and social network connecting professionals who have interacted with one another in both the personal and professional realms.

Everything Else

Social news, social bookmarking, and social networks are the main social sites at present, but they are not the only sites that allow you to share your content. Whether your passion is knowledge, photography, or video, there is an ever-growing number of websites that let you share and spread information with an audience who may already be willing to listen.

The Web As a Means of Giving Consumers a Voice

Prior to the advent of the social networking sphere, you had to have a substantial amount of money to share your content on the Web. To establish a web presence, you needed to hire a savvy web developer and a knowledgeable graphic designer. You also needed a domain name (approximately $70/year), and most importantly, web hosting space. That's why, until the turn of the century, the only professional websites were ones owned by companies. Few individuals really had their own personal web spaces.

In the past several years, however, numerous things have changed. First, social sites emerged and gave us the ability to create our personal space on the Internet. Not only that, those social sites let us connect with others who had common interests or backgrounds.

Second, social web applications have become increasingly popular. In the previous section, I described some of these applications that are hosted by various companies and give users the ability to set up their own profiles and establish relationships. Additionally, open source applications (such as MovableType, WordPress, Drupal, and Joomla) allow individuals to become their own publishers. These applications have evolved substantially in the last few years and can be set up by just about anyone with an Internet connection, as they do not require extensive technical know-how. Savvy web developers and knowledgeable graphic designers are still being sought out, but some applications exist that eliminate the need for some web designers, and there are thousands of open source (and cheap) themes that eliminate the need for a costly graphic designer.

Third, domain names have become more easily accessible and affordable. The dot-com domain names are likely all taken (unless you can think of something imaginative that nobody else has considered), but new domain name extensions are being released on a fairly consistent basis, with the recent launch of .me and .tv domains. The more popular domain name extensions now cost between $7 and $10 per year.

Fourth, web hosting has also become a lot cheaper. If you want to create and manage your own personal space, you can download the open source applications and install them quickly on your web host. Whereas hosting may have cost a few hundred dollars before 2000 (and data delivery was a lot slower), now even children have their own web spaces and the cost is a fraction of what it was in 1999, with significantly faster speeds.

Before social media, you heard about new products through traditional forms of marketing: newspapers, magazines, or perhaps television commercials. Those tried and true tactics are no longer as powerful as they once appeared to be. Before social media, you would use the same traditional media channels to read about a bad company or a great one from consumers who had firsthand experience. Nowadays, the picture has changed.

With cheaper and faster technology, the Internet has evolved into a different kind of beast. Moreover, the extent of communications now travels farther than imaginable. Performing a search using your favorite search engine indicates just how far we have come. In 2001, a search for Comcast on Google yielded the results shown in Figure 1-2.

Today, the same Google search doesn't give you quite the same information, and that information is no longer controlled by a single entity (see Figure 1-3).

As shown on the search results page for the 2008 search, the expected results for "Comcast" appear—there are numerous links for the company's website and its various divisions (and for viewing the company's financial information), as well as news stories related to Comcast. Further down, however, you see social media finding its way to the top of the search results. First, you can see Wikipedia, the user-generated encyclopedia of information about public companies and notable individuals, among other informative topics. Perhaps more alarmingly, a potentially troublesome YouTube clip has found its way to the first page of the search results.

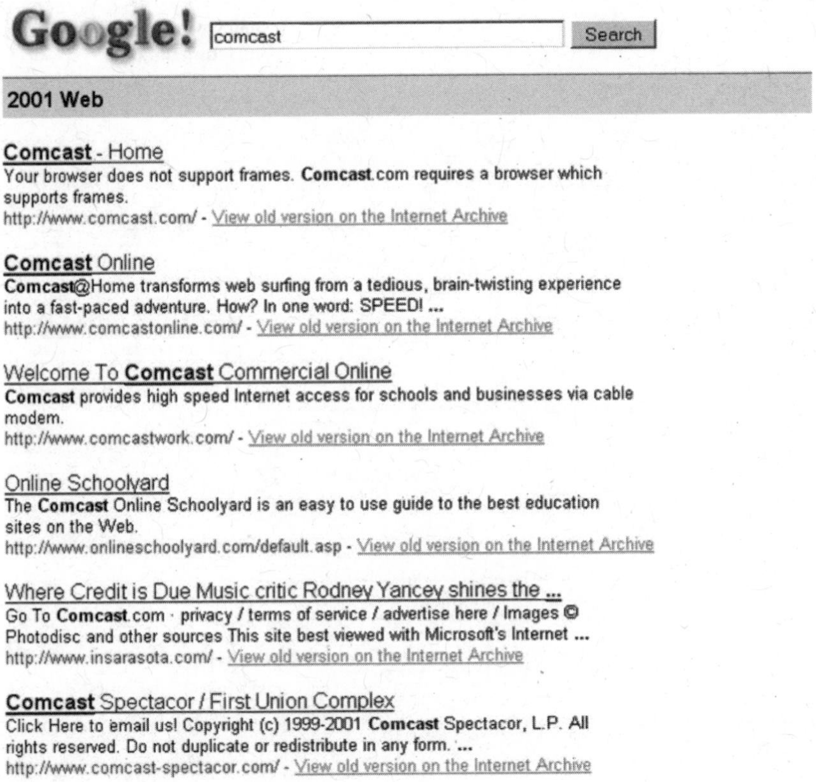

Google! | comcast | Search |

2001 Web

Comcast - Home
Your browser does not support frames. **Comcast**.com requires a browser which
supports frames.
http://www.comcast.com/ - View old version on the Internet Archive

Comcast Online
Comcast@Home transforms web surfing from a tedious, brain-twisting experience
into a fast-paced adventure. How? In one word: SPEED! ...
http://www.comcastonline.com/ - View old version on the Internet Archive

Welcome To **Comcast** Commercial Online
Comcast provides high speed Internet access for schools and businesses via cable
modem.
http://www.comcastwork.com/ - View old version on the Internet Archive

Online Schoolyard
The **Comcast** Online Schoolyard is an easy to use guide to the best education
sites on the Web.
http://www.onlineschoolyard.com/default.asp - View old version on the Internet Archive

Where Credit is Due Music critic Rodney Yancey shines the ...
Go To **Comcast**.com · privacy / terms of service / advertise here / Images ©
Photodisc and other sources This site best viewed with Microsoft's Internet ...
http://www.insarasota.com/ - View old version on the Internet Archive

Comcast Spectacor / First Union Complex
Click Here to email us! Copyright (c) 1999-2001 **Comcast** Spectacor, L.P. All
rights reserved. Do not duplicate or redistribute in any form. ...
http://www.comcast-spectacor.com/ - View old version on the Internet Archive

FIGURE 1-2. 2001 Google search results for "Comcast"

The Wikipedia and YouTube social sites have beat out over 34.7 million pages to appear on the
first page for a very simple and common web search.

Now that web technology is cheaper and content is produced more easily, the Web has become
a means of giving consumers a voice, and as shown in the Comcast example in Figure 1-3, that
voice may not be in your company's favor. In 2001, it was unheard of to start a blog and
complain that you were unhappy with the delivery of goods. Only recently have websites
emerged that are dedicated to letting consumers fight back or rant about poor service. Blogs
are now media for complaints about products and services (among other innocuous ramblings).
Reputation management, or responding to negative mentions of your company or product on
the Web, has blossomed as an industry within the social media marketing/search engine
marketing sphere.

There appears to be no shortage of negative press about products (there is positive press, too,
but as many of us know, the good stuff simply doesn't spread as much as the bad stuff does),
and the Internet has allowed consumers to seek reparations for injustices and to ramble about

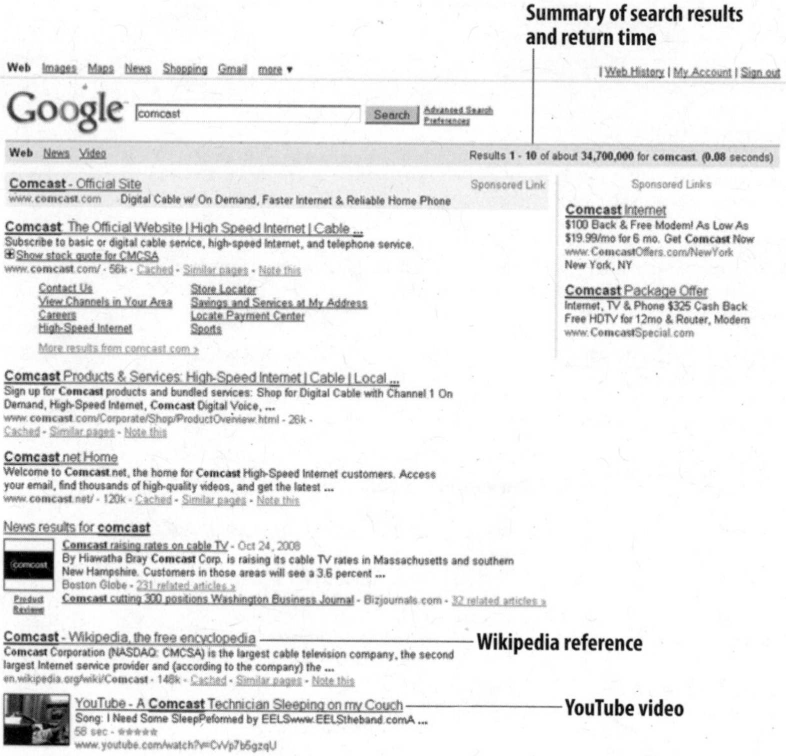

Summary of search results and return time

Wikipedia reference

YouTube video

FIGURE 1-3. 2008 Google search results for "Comcast"

their dislike of a particular product. Since negative publicity is often most talked about, this also means that people are sharing these stories and likely linking to them as well. The more links to a story, the higher the stories are (usually) ranked in the search results. That's why you see a video of a sleeping Comcast employee on the front page of the Google search results. If you look at the whole picture and the entire search results for Comcast on that particular results page, that video is probably the most interesting link.

This proves that the Web is giving everyone the opportunity to share whatever they deem relevant to their audience. It's easy to set up an account and start criticizing the editor of your newspaper or to openly disapprove of the way your school board is handling discipline problems within your district. Sometimes a single blog post, if ranked well (and discussed enough), can adversely affect your business, especially since consumers are often seeking out reviews of companies before taking the plunge. If that negative press is highly visible, it's likely that consumers looking to make a purchase decision may look to a competitor who isn't facing bad press.

It's Time to Join the Conversation

What do you do when you discover that someone is speaking ill of your company on her website or on another web channel that is accessible by the general public? The traditional approach would be to sit back and wait for the wave of complaints to stop. Today, with the ease of spreading information, that is not the ideal approach. Instead, joining in the conversation may be the best route you can take.

In the past, consumers would simply soak up what they read in print media and what they watched on commercials. Consumers had a limited floor upon which to give feedback to the message providers. We called this a monologue. This old way of communicating has changed, however. The Internet has facilitated two-way conversation. Conversations about your product are happening online regardless of whether or not you are participating in them.

Marketers have the responsibility to be ahead of the curve and to pay attention to these conversations. They should understand how individuals perceive companies and products online, and they should engage in a fully transparent dialogue with openness and honesty. The marketer who speaks to those who will listen and will actually engage in the two-way dialogue on a consistent basis—on matters good and bad—is the marketer who will be able to build trust and drive those sought-out conversions.

Take the lead and don't languish in the echo chamber. Seize the opportunity and join the conversation. Consumers in this digital era appreciate transparency and communicative entities, and it's never too late to communicate with your constituents. They're waiting.

Content Is Not King (Not by Itself, at Least)

There's a very common saying in the search engine marketing sphere: *content is king*.

In other words, you can write everything and your visitors will come. You can start a blog and just keep writing. Eventually, people will find you and become loyal followers. You can write an exhaustive article on a topic about which you are incredibly knowledgeable, and hope that the right people eventually discover that article and share it with their friends. That is the meaning of the "content is king" mantra.

This idea by itself, however, is not entirely true. How many times have you accidentally stumbled on a great article online that nobody seems to be talking about? Chances are, it happens more often than not.

Gary Vaynerchuk, host of the popular online TV show Wine Library, uses a more appropriate slogan: *content is king, but marketing is queen (and the queen rules the household)*.

The bottom line is that you can't just write something and let it sit there. How will people find your content if you don't disperse it appropriately to the right communities, or if you don't share it with anyone at all? The key takeaway is that marketing is part of the mix, and if you are producing content in the online realm, it is crucial to leverage the social media community to spread that message and share the information with the individuals who will absorb it and pass it on.

I'll end this section with another important quote, from Michael Gray, a search engine optimization expert:

> Building quality content without [marketing] is like locking William Shakespeare in a room to write for himself.

Are You Ready for Social Media Marketing?

You've read this far and are intrigued by the benefits of social media marketing. Yet are you ready to take the plunge into a vastly different communication tactic? Some companies are not at all prepared, and some simply won't ever achieve the success that they are hoping for. The reason for this is that they are not willing to give up control of the conversation. They are afraid that when they introduce the community into the discussion, they may not hear what they want to hear, and their responses (or lack thereof) can distort the public's perception even further. Unfortunately for those who are unwilling to innovate and dedicate resources to these peer-to-peer channels, social media is here to stay.

There are two main considerations when assessing your readiness to embrace social media marketing.

You Must Be Willing to Give Up Control of the Message

Today, everyone can be a content creator. After all, there are hundreds of thousands of websites on which individuals can be publishers, and it takes little to no effort. People are using these sites to talk about you.

Companies need to acknowledge that they can no longer easily control their messages. Marketers now have the ability to influence and cultivate the message through their own communication channels and through regular community participation, but they are now contending with hundreds of thousands of customers who have a soundboard to articulate their own thoughts about the company and product offerings. Marketers should listen—not ignore—these messages, as they can provide some deep insights into the presentation of the product and the actual marketing message, and companies may find suggestions on how to improve.

You Must Be Willing to Dedicate Time and Energy to Achieve These Goals

In the digital space, the truth of the matter is that not everybody gets it. Getting it doesn't happen overnight, either. It will be necessary to allocate the resources to achieve the goals. This will often require manpower.

It's also important to note that the initial time commitment will probably be substantial. You'll have to study communities, learn the proper rules of etiquette (some sites may demand different rules than others), and figure out how to respond based on what is acceptable within the community. Over time, as you gain knowledge, the time commitment will be less significant, but you'll still have to keep abreast of developments in the field in order to stay ahead of the curve. Building and maintaining trust will require a consistent time commitment, as you reinforce your involvement in the community and remind your constituents that you are a participant in the community for the long haul.

So What's Next?

There are online conversations about your company, product, or service going on *right now*, and they will happen regardless of your participation. It is your responsibility as a marketer to find out exactly what people are saying and how they perceive you. By becoming involved, you can facilitate that conversation, sway your audience, and engage community participants in a dialogue that will be beneficial to both them and the entity that you represent. Such an engagement can translate into tremendous successes for your marketing message, from reputation management to increased brand awareness, and then some. What are you waiting for?

Summary

Social media usage is on the rise with billions of frequent online interactions, and as such, social media marketing is a great way to connect consumers with companies and brands.

Social media marketing is about listening to and sharing great content with the collective. This helps drive links, raise brand awareness, increase conversions, and kick-start conversations. This is a much more powerful tactic than the old practice of traditional advertising; the old strategies are no longer as effective.

One of the biggest challenges of social media marketing is measuring your ROI. Much of social media is not that easy to measure; you can't put a numeric value on buzz and quality of conversation.

Social media portals, where conversations are ongoing and marketing messages can be conveyed (if done right), exist in all across the Internet. We reviewed social news sites, social bookmarking sites, and social networks in this chapter, but many other portals exist as well.

Social media is making its way into search results, and you have the opportunity to join in on the conversation. It's not in the best interests of your company to sit back and let that conversation continue without stepping in to influence it.

You may have a great product or an excellent whitepaper that will help boost your brand, but without marketing it via social media, nobody will discover it. After all, content is *not* king (not by itself, at least).

To be sure you're ready for social media marketing, you will have to give up control of the message and dedicate time for the task. In later chapters, we'll look at which strategies are most effective.

Endnotes

1. *http://www.universalmccann.com/Assets/UM%20Wave%203%20Final_20080505110444.pdf*

2. *http://www.w3.org/People/Berners-Lee/ShortHistory.html*

3. *http://www.rohitbhargava.typepad.com/weblog/2006/08/5_rules_of_soci.html*

4. *http://www.socialmediaexplorer.com/2008/10/28/what-is-the-roi-for-social-media*

5. *http://www.readwriteweb.com/archives/digg_the_decline_and_fall_of_tech.php*

6. *http://www.soshable.com/reddit-changes*

CHAPTER TWO

Goal Setting in a Social Environment

Before you embark on a social media marketing campaign, you need to clearly articulate and understand exactly what it is that you intend to achieve. What are you hoping to accomplish? Are you looking for more eyeballs, conversions, or both? In this chapter, we'll review the essential steps required to develop your social media marketing campaign and offer solutions for a variety of scenarios that such a campaign can address.

It's important to acknowledge that social media marketing, without goal setting, can turn out badly. Social media marketing is about real, personal relationships. As such, you must listen and respond appropriately. It takes research and careful planning to determine how you will engage with community constituents. If you are jumping into the game without being aware of your surroundings and your space, the consequences can be disastrous. For example, most homeowners are not very appreciative of telemarketer intrusions during dinner. In the online space, the sentiment is similar. People can choose who to listen to and who to weed out. If your mission is to "sell, sell, sell," and not to "give, give, give," you will fail because people are no longer receptive to forceful marketing tactics. Clearly outlining the goals of your campaign and then carefully reviewing the steps that are required to achieve these goals will give you a better grasp of your needs and how to attain them.

The Hurdle: Overcoming Fear About an Uncontrolled Message

In Chapter 1, we briefly discussed the biggest fear of companies and brands as it relates to social media and social media marketing: *giving up control of the message*. In traditional media, the conversation was one-way. You spoke, and an audience listened. Today, the communications climate has drastically altered: we're facing a medium that is composed of millions of people who can actually contribute to or detract from a marketing message. Thus, social media marketing is inherently *social*. It is a two-way conversation. Brands, marketers, and companies talk, but this time, the audience is part of the discussion. There is a balance of power between you (the representative of the brand) and the people (those representing the market).

Consider a site like Amazon.com. Amazon offers millions of products from books to home repair products, and its marketplace is very interactive. A popular product can easily have thousands of reviews. Clearly, as Figure 2-1 shows, brands and products are heavily discussed on a consumer-to-consumer level.

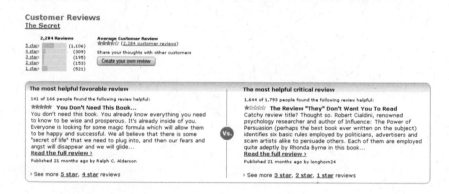

FIGURE 2-1. Thousands of customers review web products

Yet there are hundreds of thousands of ongoing conversations elsewhere on the Internet. Where are these conversations occurring? As this book suggests, websites that allow users to create their own virtual spaces can foster and ignite these discussions. A customer who is disgruntled about poor service and support (poor package handling by a shipping company, disappointment over a contractor who did not deliver what was promised, or anger over a restaurant that undercooked an expensive ribeye steak, for example) could easily launch a blog where he will, if he wishes, discuss his dissatisfaction and possibly use the personal web space as a grounds to tear apart the reputation of the company. On the other hand, satisfied customers have also launched web pages and uploaded videos that show enthusiasm for products they have purchased or services they benefit from. The conversations can go both ways to express both positive and negative emotions. Standalone and noninteractive pages can crop up and contribute to the conversation. Blogs often showcase positive (and negative) experiences from companies and products. With social media monitoring and engagement, access to this sentiment and information is available to businesses for free.

THE POSITIVE (BUT MOSTLY THE NEGATIVE):
WHERE TO READ (HORROR) STORIES

Numerous websites exist solely to highlight companies in a positive or negative light. The sole purpose of these sites is often to share stories about interactions with particular companies and to create a connection between the consumer and the brand. Some of these sites include:

Get Satisfaction
> Get Satisfaction (*http://www.getsatisfaction.com*) is a people-powered customer service portal. It offers to connect customers with the employees of the companies they're concerned about. Notable companies on Get Satisfaction include O'Reilly, Adobe, Whole Foods, Timbuk2, Apple, BBC, and Mozilla.

The Better Business Bureau Complaints Portal
> Not so much a "social media" site as we know it, the BBB (*http://www.bbb.org*) takes complaints about businesses both online and offline, which are accessible by the general public via a site search.

Ripoff Report
> ROR (*http://www.ripoffreport.com*) lets disgruntled consumers share their stories with other disgruntled consumers.

Yelp and Merchant Circle
> Yelp (*http://www.yelp.com*) and Merchant Circle (*http://www.merchantcircle.com*) both allow real people to write reviews about businesses that they've had personal experience with.

Consumerist
> Consumerist (*http://www.consumerist.com*) is a consumer affairs blog that highlights stories, both positive and negative, about consumers' experiences with companies and corporations.

Consumers now have a real voice on the Internet, and that voice is becoming louder. With most of the aforementioned tools, unhappy customers already have forums to share their feelings with communities of individuals in similar situations who can relate and offer advice. They can use any of the these sites to kick things off, or they can use their own web spaces and other social channels to alert the entire world to their sentiments.

There are numerous tools that you can use to monitor these conversations, and we will explore them later in this book.

The new media mindset emphasizing two-way communication is critical: it's important to have a conversation *with* someone, rather than speaking *to* someone. In fact, it really can be a win-win situation for anyone who opts to engage in social media marketing. However, those looking to proceed with a social media marketing model must employ other tactics as well.

Transparency Is Critical

Social media marketing is all about disclosure, or transparency. If you educate your audience about your primary goals and core values, and let your constituents know exactly what you are doing right and wrong in your company, you will have nothing to fear about social media. Being upfront about your objectives is becoming increasingly important, as suggested by the *Times*, which stated that "fake bloggers will be named and shamed."[1] If you are not upfront throughout the entire duration of your campaign, someone may eventually find out your true objectives and you'll have a lot of cleaning up to do to reestablish your online reputation. Consider social media marketing your modern-day public relations.

LACK OF TRANSPARENCY: EDELMAN'S WAL-MART CAMPAIGN

In 2006, Wal-Mart launched a blog called "Wal-Marting Across America." The premise behind this blog was to chronicle the journey of two working-class Americans who were traveling the country and spending the night in Wal-Mart parking lots.

The problem? The blog was not a product of Wal-Mart—not entirely, that is. It was, in actuality, a fake blog orchestrated by Wal-Mart's public relations firm, Edelman.

When the identities of the bloggers behind "Wal-Marting Across America" were discovered, there was a public relations fiasco for the involved parties. Bloggers and news outlets everywhere expressed their distrust in the participants and cited concerns regarding ethical word-of-mouth marketing.[2] The two bloggers who were hired for this Wal-Mart promotional tactic were initially under attack and intense scrutiny for their involvement, but at the end of the day, those blamed for the failed mission were Wal-Mart and Edelman.

Today, "Wal-Marting Across America" no longer exists.

Being upfront with your constituents is incredibly frightening if you are unprepared. However, full disclosure has its benefits. If you make a business decision, it's important to communicate its upsides and downsides. If you make a mistake, it's important to own up to it and to let your customers know that you are putting them first. Keeping the floor open for comments and feedback and responding in kind is what social media marketing is all about.

OPENING UP ABOUT YOUR MISSTEPS: JETBLUE'S PROMISE TO ITS CUSTOMERS

Valentine's Day 2007 was a bad time for JetBlue. Weather conditions caused flights to be cancelled, then rescheduled, and then cancelled again. For some flights that weren't canceled outright, the airline instead left passengers stranded on planes for hours. This operational crisis resulted in negative publicity and lost trust from its stakeholders.

JetBlue could have remained silent, but instead it acknowledged publicly that it had wronged its customers. CEO David Neeleman said he was mortified. JetBlue published full-page online and print advertisements saying, "We are sorry and embarrassed. But most of all, we are deeply sorry."

The apology continued in a YouTube video in which Neeleman spoke directly to his stakeholders with an apology that pundits have agreed was authentic.[3]

The lesson learned is that companies should not sit on their mistakes. Transparency means being open from the get-go, but also acknowledging if that you've committed a sin and sincerely promising that you will never commit it again.

To date, JetBlue has made good on its promise.

Listening Is Required

A two-way conversation means that you, as a marketer, are spreading your own business message, but you are also listening to what is being said about your products and messages. Listening itself is not enough, since your presence may go unnoticed; by responding, you reinforce the relationship with your audience and let it know that you value its insights. A successful social media marketer maintains accounts on all social media platforms and responds when appropriate. Yoono, a browser-powered social network, has preached the importance of social media engagement across several networks. Regan Fletcher, Vice President of Business Development at Yoono, said the following in an interview with me about the importance of listening:

> For a startup developing a consumer tool, the most important resource we can tap into is feedback from users and potential users, even when that feedback is negative. Especially when that feedback is negative! Generating feedback from different social media outlets has helped Yoono to launch a consumer tool without the high cost of formal focus groups. We still plan to organize a real focus group in the future, but that will come after we've implemented feedback from users who follow us on our blog, Twitter, Facebook, or FriendFeed. And when we do actually sit down to plan a true focus group, we'll use social media to help organize and attract participants instead of hiring someone to do it for us. We actually did contact a couple companies that organize focus groups, and the first quote we received was $230,000! I've used the

expression "my jaw dropped" many times before, but this time, it literally did. That kind of money supports our entire company (salaries, office space in two countries, servers, travel) for more than two months!

Once we saw some of the user feedback generated on Twitter and Facebook, we looked for ways we could expand beyond our current users. We organized a series of Social Media Camps where users could talk in BarCamp-style (that is, user-generated-conference-style) about how they were marketing with social media. We learned a lot from these events, but the very first lesson came a month before the first event and could not have happened in the pre–social media world. We designed a logo for Social Media Camp, put it up on Flickr, and invited feedback. Right away, people told us it was too corporate and strayed from the spirit of BarCamp. We made some changes and came up with something that everyone liked, but more importantly, showed our community that we were open to their feedback and willing to act upon it.

Regardless of whether or not you choose to embark on a social media marketing campaign, you will be talked about. It's better to engage in a two-way dialogue that will provide you with insights into your constituency and feedback on ways to improve. And as Yoono showed, the more active you are within the social media communities, the greater the chances you have to develop genuine relationships with your target customers—and then some.

THE COST OF SILENCE: THE DELL CASE STUDY

As early as 2005, social media began to have an impact on customer relations. When his laptop malfunctioned, influential blogger Jeff Jarvis used his blog to express his dissatisfaction with Dell's "appalling" customer service. After several blog posts, Dell remained unresponsive to his multiple requests for help, and Jarvis was frustrated. Finally, he wrote an open letter to the company's CEO on his blog.[4] His blog post had over 10,000 views in a short period of time, and to date, has over 700 comments, many from individuals who also felt like they had received poor customer support from the PC manufacturer. After the media picked up on Jarvis's rant, Dell made an effort to contact him and ultimately issued him a refund for his problematic machine.

Most significant, however, was Dell's follow-up response and how it ultimately showed that the company was nurturing bloggers and social media participants rather than ignoring them. After the Jarvis incident, Dell decided to launch its Direct2Dell blog, which gave the company a human voice. In February of 2007, Dell launched IdeaStorm.com to solicit advice and feedback from Dell computer owners. Dell leverages user input on the site to determine what customers are seeking from Dell products.

And, of course, Dell is still following bloggers. In a follow-up article by *BusinessWeek*,[5] Jarvis acknowledged that negative blog posts about Dell have decreased from 49% to 22%.

Asking the Right Questions: Setting Goals for Your Campaign

Before you actually embark on your social media marketing campaign, you must clearly define your goals by asking yourself what you are trying to achieve. Some social media marketing campaigns arise from the need to push negative results out of the search engine results pages (see the Comcast example in Chapter 1). In this case, you are aiming for two different goals: reputation management and new links that will emphasize other relevant pages in the search engine results.

Let's discuss the various goals of a potential social media campaign and review scenarios where these goals can be applied.

There are several possible widely known social media marketing objectives that a campaign can allow you to achieve. Later in this chapter, we will get more granular in our goal-setting methods.

Increased Traffic

One of the primary objectives for many social media marketing campaigns is to increase traffic, either among a relevant audience, or in a general effort to gain more eyeballs with the hopes that a fraction of them will be relevant. Social media campaigns of this type may focus on generic social sites or pitch to bloggers who cover a wide variety of topics. Increased traffic can thus be either highly relevant or not relevant at all. However, this may be the exact goal of the social media marketing campaign.

What does increased traffic do for you?

Increased traffic typically translates to higher rankings on sites like Alexa or Quantcast, both popular site-ranking tools. This can result in raised prices for advertising costs if marketers and content creators aim to monetize their own content. A site that boasts 5 million monthly unique page views can easily make the argument that its ads should cost $5,000 versus a site that only achieves 50 unique visitors a month.

Because increased traffic often correlates with increased trust, other opportunities are available as well. Increased traffic can aid in the success of other social media objectives, such as brand awareness, reputation management, and improved search rankings.

There's just one catch...

When the traffic boost to the website is not relevant, bounce rates are very high. In many instances, visitors are not interacting with the website and will only be interested in the content being shared directly through social channels. Sometimes, this can be attributed to the lack of a call for action. Most importantly, the social media marketing campaign should aim to increase traffic that is relevant. For example, at a recent conference I attended, a disability website marketing director complained that her extremely social media–friendly content, which earned hundreds of thousands of visitors in a short period of time via social news site Digg, was not converting into real long-term visitors who would stay on the site and engage with the actual disability site subject matter. She was upset that she invested in a "great blogger" who provided valuable content, but that the site's visitors would consume the social-media friendly content and then leave shortly thereafter. One sample piece of content was related to photographs of funky urinals. The urinals had absolutely no relevancy to the mission or goal of the disability website, and naturally, visitors did not venture beyond the Digg submission.

DEFINITION

A *bounce rate* is the percentage of visitors who enter and exit a website on the same page without visiting any other pages on the site in between. A visitor who contributes to the bounce rate goes to the linked-to web page and immediately exits without any site engagement.

When crafting a social media marketing message, if your goals are to get targeted traffic and maintain a percentage of your visitors and convert them into regular community participants who frequently engage on the site, you must ensure that you maintain some relevancy. Tie in your social media marketing message with the mission of the site. While social media crowds love the zany and unique content, if the content itself is not at all related to your site's core values, your bounce rate will be substantially high.

Increased Brand Awareness

If executed properly, social media marketing can have a profound effect on brand awareness. A study performed by Immediate Future[6] found that a brand's engagement in the social media sphere has a direct correlation with the reputation of the brand. Immediate Future found that popular brands have very heavy involvement in social media spheres, though there is not one social network or facilitator that enables this communication. Engagement across several social networks has bolstered brand awareness.

When companies choose to embrace social media enthusiasts, another unique shift occurs. Brands that embrace individuals who speak both positively *and* negatively of products and services, and that indicate that they are actually *listening* to their constituents, have seen another paradigm: they have the power and ability to turn those individuals into brand ambassadors, individuals who can positively advocate the brand.

A *brand ambassador* or *brand evangelist* is an incredibly powerful tool for social media marketing efforts. In general, when individuals consider buying products or services, they will perform initial research online before they choose to purchase or abandon the conversion process. Consumers often are resistant to the marketing message put out by companies, but they are not as resistant to recommendations and stories from their peers. They would much rather hear from real people than from individuals who are clearly representing the company.

The trust is simply not there. Empowering brand ambassadors can be extremely valuable to your company.

Improved Search Engine Rankings

A successful social media marketing campaign can generate hundreds to thousands of links because viewers will want to share the campaign's web page with their friends, family, or—if they're web influencers—their audience. Essentially, social media marketing is a modern-day link -building strategy that encourages creativity and keeps you from having to hunt for the right webmaster to link to the content. While search engines' algorithms are not straightforward, one thing is certain: the more links that are pointing to your pages, the more likely you are to raise your search engine rankings.

Reputation Management

Somewhat related to improved search engine rankings is the impact of social media on reputation management. Social media can aid reputation management in many ways (and will be explored in more detail in Chapter 4). It's apparent in the Comcast example in Chapter 1 that social media sites are visible in the search engines, and these pages can be ranked quite highly. Quite the same way, you should take a proactive stance toward social media to promote your company and brand. Why? Social sites are given a measure of trust by search engines, and by creating your own company profile on these sites, you can have a presence in search engines that may help push down the negative results. Further, great content that a lot of people link to (often utilizing social media methodologies) can also push negative search results so that they are less visible.

Search engine rankings are only part of the equation, however. Companies can use reputation management tactics to avert publicity disasters by turning negative experiences into very positive ones. With social media monitoring and conversation tracking, companies can turn negative past incidents into positive growth experiences for companies and their brands. In fact, as we will see in several case studies later in this book, some companies have transformed what could have been serious reputation issues into incredibly powerful marketing messages.

Increased Sales in Your Product or Service

Social media tactics, such as user-generated reviews and video tours, can increase sales in products or services. Several studies have been conducted over the past few years to support this claim. Among them, these studies from 2007 and 2008 show that social media does have an impact on increased sales:

- A 2008 Nuance Communications study performed by the Society for New Communications Research for[7] shows that a hefty number (nearly 75%) of study participants purchased products and retailers based on peer recommendations. The influential social media sites referenced in the study included discussion forums, blogs, and online rating systems.

- A SellPoint study conducted by Coremetrics[8] found that users were more likely to buy products after viewing video tours, translating to a 35% increase in sales. The study tested over 1 million shoppers' behavior on the CompUSA website over a 60-day period.

When individuals are looking to purchase a particular product, they will likely turn to the Internet and read the reviews from their peers before they take the plunge. Of course, improved brand awareness can contribute to higher conversions as well.

Established Thought Leadership

If you possess knowledge that individuals are seeking out, social media marketing can work in your favor. Social media engagement, especially though channels that encourage the sharing of ideas and tools, can help you establish yourself as an expert. Many bloggers who have been writing on a regular basis have found themselves in a higher status online due to the ubiquitous reach of the Internet: they have become established subject-matter experts who are able to educate not only their own customers, but also peers who seek out their advice. As such, individuals who are able to establish themselves as thought leaders can forge new friendships and build new business relationships both through new customers and other "experts" in the field.

> ### DEFINITION
> A *thought leader* is an individual whose knowledge and expertise has positioned him as an expert among his peers.

Social Media Marketing Scenarios

Social media, then, can do many things: it can increase traffic to your site, increase brand awareness, improve your search engine rankings, act as a reputation management tool, increase sales in your product and service, and help you establish yourself as an authority.

Now let's outline numerous scenarios that can benefit from social media marketing to some extent. These examples are just a few possible implementations of social media marketing campaigns.

Scenario: An online publication seeks additional page views and traffic

In the scenario where online publications, particularly those dependent upon CPM advertising, look for traffic and page views, social media is an incredibly viable tactic. After all, by writing the right piece of content on the variety of already existing social media channels, you can easily get hundreds of thousands of page views. In a particular campaign run in August 2008, a popular publication received a 353% increase in visitors by utilizing social channels. The strategy entailed researching the social sites that would be receptive to the content, submitting the content to them, and sharing the submission directly with individuals who were likely to pass the articles on to their peers. This, in turn, became an exercise in viral marketing. The most successful of the social submissions received tens of thousands of unique visitors to the original article in a short period of time and consequently garnered hundreds of links and bookmarks.

It's important to note that not all content will work on social media channels. Finding the appropriate content, packaging it correctly so that social media crowds will appreciate how it is displayed, and then submitting it to the right social sites with a good title, description, and tags is very important.

Scenario: You have a product and want to get the word out

There are many different social media marketing approaches that may work with new product offerings. One example is blogger outreach. In 2005, Andrew Milligan from beanbag company Sumo Lounge (*http://www.sumolounge.com*) was in a marketing quandary. He had a great product—versatile beanbag chairs—but didn't have the marketing budget to pay for ads. Instead, he contacted numerous blogs and asked if they would consider reviewing his products on their sites. Over a period of three years, he sent beanbag chairs to numerous bloggers who then used their personal web spaces to talk about the product. As a result, Milligan saw an initial increase of 500% in sales with a positive ROI from year to year. Sumo Lounge, which was once insolvent, became a multimillion-dollar company with success that can be attributed directly to reaching out to bloggers and having them review the product. Milligan says, "Altogether it was a lot of work but it was well worth it and paid off in the end."

Other strategies, such as creating videos, can yield product awareness. In May 2008, a video titled "Why Every Guy Should Buy Their Girlfriend a Wii Fit"[9] was uploaded to YouTube. The video shows a twenty-something male secretly videotaping his girlfriend playing the hula-hoop game with the Wii Fit. That video now has over 8.3 million views. On the same day it was posted to YouTube, the video was submitted to social news site Digg, where it now has more than 11,000 votes. What's more, the original video then spawned over 35 spoof videos, which together have amassed millions of views.

CASE STUDY: BLENDTEC

In 2006, George Wright, marketing director for Blendtec, a manufacturer for both personal-use and commercial blenders, was given a $50 marketing budget to do something original with the company's powerful but not very well-known products. One day, as Wright was walking around Blendtec's conference room, where his colleagues often demonstrated how strong the appliances were, he spotted a pile of sawdust on the floor. Later, he learned that a chunk of wood was "blended" to show prospective buyers that Blendtec's blenders were super-strong appliances.

Using his very low marketing budget, Wright bought a domain name (willitblend.com), a lab coat, a rake, and a bag of marbles. After successfully filming founder Tom Dickson blending these items, and then posting the videos to YouTube and to its branded site, willitblend.com took off. To date, approximately 75 videos have been launched illustrating the unique blending power of Blendtec blenders.

Blendtec's success did not end there. The videos have had over 185 million views across social media sites and willitblend.com, and the Will it Blend? channel is YouTube's 34th most popular channel. Additionally, sales of Blendtec products increased by a resounding 700%. Blendtec's brand became known worldwide, with media mentions in dozens of renowned publications. George Wright has been invited to speak about the company's success at trade shows around the world.

About his company's success, Wright offers solid words of wisdom: "Small companies can have a big presence. The rules have changed." He adds, "Instead of making ads, think of making content."

Scenario: A search for your company name yields negative pages listed in the first four search results

Social media allows individuals to broadcast their feelings, positive or negative, about their experiences. Often, these stories can find themselves high up in the search engine rankings and can have a profound impact on viewers' decisions to engage in the product offerings.

In a business-to-business eye-tracking study performed by MarketingSherpa and Enquiro,[10] executives were asked to view real web pages. The heatmap shown in Figure 2-2 suggests that items *above the fold* (near the top of the page) got more visibility than text and images below the fold.

Similar studies have shown that items listed higher in search engines have more visibility and more page views. How many times have you heard from your marketing department that you want to be ranked #1 in Google for your desired keywords? Often, that's a strong selling point and can indicate much success.

If negative reviews about your company are extremely visible in the search engines, your sales can be affected. Your customers, upon researching your product, may opt to go with a competitor who does not have visibly damaging search engine results. Utilizing social media avenues and taking a more proactive stance in social media—from engaging with the individuals who may be posting those damaging stories to working to create solid social media profiles and consistently working to develop them—are healthy ways to join in the conversation. Better yet, in due time, those negative results may be pushed down in favor of social media stories and profiles that indicate a positive engagement.

Reputation management is critical in social media because you can manage relationships, but you can't manage reputation. However, the two are related. Successful relationships can

FIGURE 2-2. Items "above the fold" receive more visibility

strengthen your reputation. Reaching out to individuals who speak both positively and negatively of your company can have incredibly powerful returns.

But why should you speak with someone who talks negatively about your company? Individuals who post negatively about your product are those who are passionate enough about it to share their dissatisfaction with their readers; they also are looking for people willing and ready to listen. They felt strongly enough that they complained. By reaching out to them and empowering them with special privileges, you motivate them to become more passionate about your brand—this time, in a more positive light—and you may ultimately turn these individuals into the most unlikely group: brand evangelists. Consider this logic: complainers are talking about your brand anyway, so why not engage them to speak about your brand in a more positive light? It's amazing what talking with people can do.

Scenario: You're an expert on a particular subject and want to share your advice with the world

You're passionate about a particular topic:

- You went to law school and have the degree.
- You're a science whiz.
- Your business knowledge by far surpasses that of your peers.

- You are a great cook.
- You're a brilliant photographer whose work has been sought out locally.

Let's face it: if you fit any of these categories, you have knowledge that other individuals may be aching to find. They're leveraging the Internet to seek out advice, and they're searching for specific issues that you may be able to address directly (or perhaps that you've already answered). This is why you should consider starting a blog. By using a blog to offer free legal advice, science theorems, tips on business, free recipes, or photo techniques, you can establish yourself as an expert on a specific topic and by far exceed your current local geographic reach, as leveraging the Internet can have worldwide exposure. Further, you can gain additional opportunities by consistently updating your blog: established bloggers are asked to speak at trade shows and conferences, their work is featured in books, they are contacted by journalists for media stories, and they are getting a lot of new business opportunities. For example, food blogger Brown Eyed Baker (*http://www.browneyedbaker.com*) used blogging to kick-start a real baking business. Blogger Chelle says,

> While my clients thus far have been local (and my blog following international), having my blog has definitely established me as an "expert" in the baking area and people are able to see that it's just not a "fly by night" business venture, but rather an interest and competency that I have been developing for quite some time now. I don't think I would have ever ventured out into a start-up business had I not been blogging for almost two years now.

Blogging can give a struggling company much-needed visibility. It can also put employees who are authorities on a subject in a position to represent their company in a way previously thought to be impossible.

Making Your Goals SMART

You've reviewed the scenarios earlier in this chapter and are considering a long-term cross-network engagement. Now you're ready to formulate your strategy, but you also need to clearly define some goals. How do you set goals that reflect what you want your brand to become? How do you set objectives that guide you in your later strategy? In marketing, your goals should be SMART: specific, measurable, attainable, realistic, and timely. The following sections will lead you through the process of SMART goal setting.

Specific

Clearly define what you need to do. Your objectives should be specific so that you will know exactly how (and if) you've achieved your goal. The scenarios discussed earlier in this chapter should give you an idea of specific goals that you can set for your social media marketing campaign. In social media marketing, your goal of "getting new subscribers" may not be specific enough; instead, you may want to settle on a desired number of subscribers. Your desire to get

1,000 Twitter followers is more specific, but at the same time, you may want to ensure that those followers are relevant to your industry.

Measurable

You can't manage something that you can't measure, and you should establish concrete criteria for measurement. You can establish a benchmark for your desired goal and aim to achieve it over time. If your goal is to get increased page views, you may want to analyze your hosting statistics on a regular basis. If you are looking for 200 more LinkedIn connections, you should know exactly how many connections you have to start.

Attainable

Your goals may be lofty, but they should also be attainable. If you have only earned 500 subscribers to your e-zine over the course of 5 years, a goal of 500,000 subscribers in 5 months is probably not attainable (nor is it realistic). As part of setting attainable goals, you need to have a firm belief that you personally can achieve the goal.

Realistic

Can you establish these goals with the resources at your disposal? Not all attainable goals are realistic. Realistic goals look at what's available to you today, whereas attainable goals look at what may be possible. Your goals should be doable, but you should still set the bar high enough so that you will feel victory when you achieve success.

Timely

Goal setting also requires you to set deadlines for yourself. If you say, "I'm aiming for 5,000 new subscribers to my blog over the next year," you may not necessarily feel motivated to complete the task. Once the year is up, lack of motivation may cause you to push the deadline even further. Give yourself a specific milestone date. It is a lot more important for you to set a deadline for "three months from today." Now get started!

Researching Your Social Media Community

We've discussed some scenarios and looked at things you can do to market your products or services using social media marketing. However, the reality is that not every social media tactic will work with everyone. While "social" includes the collective, you won't necessarily see a flurry of men visiting kirtsy.com, a social media news site for women, and thus, your railroad and train videos and photos may not be an ideal fit for those users. Your target audience may not be on the sites you're aiming to connect with. That's not to say you shouldn't be listening to those venues for ideas, insights, and feedback, but you should not spend the majority of your efforts on sites that may not yield a substantial return.

When setting community-oriented goals for social media marketing, study the landscape and try to get an idea of the kind of content potential customers will be receptive to. To do this, ask the following questions.

Who Is Likely to Buy My Product or Participate in My Service Offering?

Are they women who knit? Men who like woodworking? Young adults who are heavily involved in technology? Of course, you will need to know exactly what kind of audience the service offering accommodates as part of your general marketing goals, and the answer to this question will help you determine your target audience. Your target audience has wants and needs outside the product offerings you provide, so it is in your best interest at this time to study these and see exactly how you can target your social media marketing campaign for maximum return. The next questions will help you assess exactly where to go from here.

What Websites Are They Visiting Online?

Among the websites your potential customers visit, are any of them social in nature? These social channels are the ones you should be targeting. You may be surprised to know how many people are blogging about subways, graphic design, automobiles, or birds, or how many forums and social communities are available for just about any specific interest. Since forums have preceded many social networks, study the behaviors of the forum members. A huge list of forums is available at *http://www.rankings.big-boards.com*. Understanding the communities where your target audience likes to congregate and understanding the rules of engagement in those communities is half the battle. For more information on community involvement and how to establish yourself as a valued member of any online community, see Chapter 4.

What Are People Saying About My Business and My Competitors?

Are people using blogs to talk about your business or your competitors? Are they using forums or social media channels to rant and rave about the aspects of the business that they hate most or the elements that they love? Wherever these community members congregate, these are the conversations you should monitor and participate in. At a minimum, you should listen in regularly on the conversation to understand the sentiment of the community. It's always important to listen before engaging. Listening will also help you formulate your strategy. You should have an understanding of the emotional dynamic before you dive into the pool, so learn to swim first. Eventually, you will be ready to respond.

What Tools and Services Does My Target Audience Use on a Regular Basis?

You may want to develop tools that will empower your users, or you might choose to write about the tools that they use regularly and how they tie in to your business. If, for example, your target audience consists of graphic designers who could benefit from applications that embed directly into the Mozilla Firefox browser, would you like to compile a list of the best tools for all graphic designers? Perhaps you may find that accountants could profit from a freeware application that is relevant to a service your business provides; this is a great opportunity for you to offer something for free, and if the tool is widely needed, you'll definitely get some attention in return!

What Kind of Content Does My Audience Prefer to Read?

The lawyer might be accustomed to reading the heavily detailed research paper. The young adult may, on the other hand, prefer more image-heavy content with funny captions and explanations. The mother-to-be may want a mixture of detailed information presented with pictures of how her baby is developing in the womb. After exploring this question and studying the social sites that feature content that was well received for particular topics, you can get a feel of the style and type of content that you should craft for your target audience.

Once you understand the answers to these questions, you'll start to get a feel for the goals you should be setting—whether you should create a blog, a video series, a podcast, an in-depth article, or a combination of any of these—and then you can find out exactly where you should focus your efforts. Bear in mind that not all social media marketing strategies will necessitate a heavy content creation strategy, though at times, compelling content is exactly what the audience is looking for.

Formulating Your Strategy

Now that you have established your goals or have a good idea of how to set them, it's time to consider an implementation strategy that will help you move forward in actually executing your social media marketing campaign. In this section, we'll discuss some other questions that you may want to consider before embarking on this journey.

Are You Ready to Handle Possible Negative Backlash?

In the beginning of this chapter, we discussed the reluctance of some companies to embark in a social media campaign because of the fear of giving up the message and then facing negative backlash. This is still something you will need to consider before jumping into the deep waters of social media.

If you don't engage, you may face negative backlash for not responding to a problem. Yet if you do engage, you may face negative backlash for doing it wrong.

Negative backlash is a tremendous obstacle for companies that still have not engaged in the social media marketing playground. But at the same time, it's important to acknowledge that positive engagement can help influence and nurture the mindset of the community and that negative backlash can turn into a positive learning experience for the company.

TIGER WOODS WALKS ON WATER: EA SPORTS GAME GLITCH?

In August of 2008, a glitch was discovered in the Electronic Arts game *Tiger Woods PGA Tour 09*. One of the players recorded the apparent software bug that showed a special "Jesus shot" involving Tiger Woods. In a video uploaded to YouTube,[11] the player showed that in the game Tiger Woods could walk on water and swing the ball on water as if it were on dry land.

This could have been perceived as a big loss for EA Sports, but instead of ignoring the message, the company decided to play along with the community. It released an incredibly clever response to the amateur video showing the real Tiger Woods walking on water.[12]

Instead of letting this glitch destroy its reputation, EA Sports decided to leverage the subculture of YouTube and proactively responded in a positive way. EA Sports successfully navigated what could have resulted in long-standing negative publicity and in turn translated it into a brilliantly executed marketing move.

The video has garnered over 3 million views, and feedback has been overwhelmingly positive.

How Will My Employees Integrate Social Media Strategy into Their Day Jobs?

Regardless of how easy it may seem, social media is a time commitment, and a substantial one at that. Do you have individuals in-house who can help you work on your marketing outreach efforts? Will you have to hire additional bodies?

> **NOTE**
>
> Recently, there have been plenty of unpaid internship positions posted on websites like Craigslist for social media outreach. Keep in mind, however, that whoever you hire should inculcate herself with your company and its culture. This job-posting strategy, while cost-effective, can be risky if the hired talent does not identify with your company culture.

Internally, you may want to designate a community manager who will take the reins on dealing with the community participants in a social media marketing project. This in itself is often a "full time" job. Additionally, you may want to consider working alongside agencies and consultants with some knowledge of the social media landscape and communities. They can help you brainstorm, facilitate outreach efforts, and craft a viral marketing strategy that will ease the transition to a full-blown social media marketing effort within your company.

You may want to outsource social media marketing tactics entirely. Some may argue that this is not ideal because of the fact that you are the most passionate advocate for your product. Hiring outside help can ease the transition, but your company and individuals within it should also seize the opportunity to work with the consultants and agencies to really convey an accurate message to the community constituents. The most ideal mix is to have the company work closely with any hired help.

Are You Willing to Take Risks and Experiment?

In any type of marketing campaign, you may have to experiment again and again before meeting with success. Social media is no different. You are still working with individuals who will react to your marketing campaign. If you fail at first, are you willing to try again? Will you be willing to grow from your failures? Are you willing to make amendments and adapt?

You can't always put a band-aid on a boo-boo. You should recognize that some elements of your strategy might not yield high performance as expected. Social media marketing is still experimental; your performance can only be assessed by the greater community. Flexibility is important. You may need to make adjustments over the duration of the campaign. You should be ready to pull out of a failed strategy if it is not faring well for your brand.

The answers to these questions will help you formulate your strategy for moving forward.

When Should I Pursue Social Media Marketing?

Social media marketing may come out of a desire to fill a specific need. For example, your product launch may be around the corner and you want your audience to know about it. As we've discussed in this chapter, a variety of scenarios may call for a social media marketing campaign.

However, there's more to social media marketing than just a campaign. It is best to be involved in a consistent dialogue and not just participate in the social media marketing playing field when damage control—or product awareness—is necessary. It is best to build up a loyal following of supporters who can stand by you when you do need a helping hand.

You should monitor conversation on a regular basis. Engagement should be the norm. You should frequently empower brand evangelists and highlight valuable initiatives for your business. Once you get a handle on social media marketing and understand how best to approach your constituents with the utmost transparency, you can participate in the conversations and foster relationships that will make your business and its offerings a lot more powerful in the eyes of your listeners.

Remember, in social media, the conversations will happen with or without you. It's best that they happen with you, regardless of whether you actually *need* to market your product.

Summary

Once you are ready to help nurture the message and have overcome the fear of a message that you will not be able to control, you need to swim in the waters of social media with complete transparency, and you must listen to the conversation.

Social media marketing can achieve numerous objectives, including increased traffic, increased brand awareness, improved search engine rankings, increased sales in your product or service, and established thought leadership. There are numerous marketing scenarios that can benefit from social media marketing. Blendtec, for example, wanted product visibility on a very low marketing budget. Today, the company produces one of the most popular video series on YouTube.

Your social media marketing goals should be SMART: specific, measurable, attainable, realistic, and timely. Don't set the bar too high, or you'll have difficulty achieving results. Before you dive into the waters, it's imperative that you carefully study the social communities and the interests of the community members. Finally, when you do formulate your strategy, make sure to account for the fact that there could be negative backlash (that can be handled gracefully), that any engagement is a substantial time commitment, and that your strategy should accommodate adjustments if necessary.

You may have goals for your social media marketing plans, but you should also consider being actively engaged on a regular basis and not just when you want something from the community. Conversations are ongoing, and it's beneficial to your company and brand for you to be part of them on a regular basis.

Endnotes

1. *http://www.timesonline.co.uk/tol/news/politics/article1361968.ece*

2. *http://www.intuitive.com/blog/edelman_screws_up_with_duplicitious_walmart_blog.html*

3. *http://www.youtube.com/watch?v=-r_PIg7EAUw*

4. *http://www.buzzmachine.com/2005/08/17/dear-mr-dell*

5. *http://www.businessweek.com/bwdaily/dnflash/content/oct2007/db20071017_277576.htm*

6. *http://www.immediatefuture.co.uk/the-top-brands-in-social-media-report-2008*

7. *http://www.emarketer.com/Article.aspx?id=1006593*

8. *http://www.marketingcharts.com/direct/video-product-tours-result-in-35-increase-in-online-sales-conversion-2491*

9. *http://www.youtube.com/watch?v=v31qxrXsxv0*

10. *http://www.marketingsherpa.com/sample.cfm?ident=30100*

11. *http://www.youtube.com/watch?v=h42UeR-f8ZA*

12. *http://www.youtube.com/watch?v=FZ1st1Vw2kY*

CHAPTER THREE

Achieving Social Media Mastery: Networking and Implementing Strategy

In the previous chapter, you learned about the questions you need to ask to devise your social media strategy. It's very important to note that all of these questions relate to *listening* and *communicating*. You need to listen and communicate consistently and regularly so that you don't lose sight of what individuals say about you.

When you actually implement your social media strategy, you will need to follow the conversation and respond where appropriate. Using the goals you have established in the previous chapter, you will need to determine which content is best for your target audience. Establishing yourself as a respected community member among your audience is an important next step. Your audience and those surrounding you will be of increasing importance when you engage in a full-fledged social media campaign.

When Is It Appropriate Not to Respond at All?

I do some occasional consulting with the Israeli Consulate, a government entity that has embraced Web 2.0 technology in many ways. The Israeli Consulate has established its presence on YouTube, Facebook, and MySpace, and it also has two separate blogs:

- One blog, IsraelPolitik (*http://www.israelpolitik.org*), discusses issues of a political nature. The blog's mission statement is "to directly address audiences throughout the world and to serve as a vehicle to better communicate the State of Israel's message of hope and peace."[1]

- In another completely separate blog, IsRealli (*http://www.isrealli.org*), anything goes, as long as topics do not touch upon the current military conflict (which is addressed and discussed on IsraelPolitik). IsRealli highlights emerging Israeli technologies, Israeli entertainment, sports, humor, and more. Naturally, if comments are directed to IsRealli that relate to the Israeli-Palestinian conflict, they will not be acknowledged on this channel.

Sometimes you will see comments that may not relate to the subject matter or require that you reveal proprietary information or disclose commentary that is not available to the public. Other comments may simply be inappropriate to respond to at all. You should feel free to comment on those topics that you feel comfortable discussing; any other topics can be ignored. If you feel that you must say something, acknowledge the feedback and comments of the individual contributing to the discussion and move on.

How Do You Monitor the Conversation?

We've established that conversations can happen just about anywhere, from sites dedicated to showcasing customer frustrations to personal websites. With the ease of publishing on user-generated websites, the sentiment expressed on these pages can negatively impact your company, and, therefore, it's important to see what people are saying about you. In reality, the discussion can happen anywhere. But it's not humanly possible to be everywhere. How can you monitor the conversation without being overwhelmed?

Thankfully, there are hundreds of tools available at your fingertips for conversation monitoring. Many are free but have a limited scope in the content they capture. Others cost money and will monitor several media at once for a monthly price. In this section, we explore some broad tools with which to monitor the conversation. The sites that follow are just a sampling of where you can find an ongoing discussion. Indeed, there are millions of websites where people may already be talking about your brand.

Free Tools

Google Alerts (http://www.google.com/alerts)
   Google Alerts provides you with the latest results discovered by Google within a variety of channels: news stories, video comments, blogs, pages found in Google's web search, and even Google's own forums/mailing lists, Google Groups. As shown in Figure 3-1, you can create a Google Alert for your company name (use quotes where appropriate). You receive the Alerts either in your inbox or your Google Reader feed.

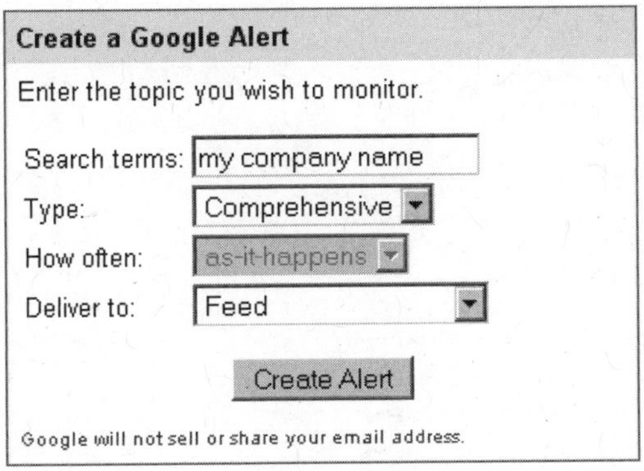

```
┌─────────────────────────────────────────────┐
│ Create a Google Alert                        │
│                                              │
│ Enter the topic you wish to monitor.         │
│                                              │
│ Search terms: │my company name      │        │
│ Type:         │ Comprehensive ▼│              │
│ How often:    │ as-it-happens ▼│              │
│ Deliver to:   │ Feed                    ▼│    │
│                                              │
│              ┌──────────────┐                │
│              │ Create Alert │                │
│              └──────────────┘                │
│ Google will not sell or share your email address. │
└─────────────────────────────────────────────┘
```

FIGURE 3-1. Creating your Google alert

TIP

Google has a number of search operators that give you more power over its search results. These work for Google Alerts as well. For example, you may want to subscribe to an alert that shows when people are linking to your company website by creating an alert for link:http://www.mycompanysite.com. People may be linking to this page without using your company name in anchor text, and it is highly recommended that you use this method to determine how people identify you.

Twitter (http://www.twitter.com)

Twitter (Figure 3-2) is a social channel, but it is also can prove to be a goldmine of information about your company. It's the perfect place for you to listen to how people are engaging with your brand and how they feel about you as a business entity. Use Twitter's official search engine (*http://www.search.twitter.com*) for real-time tracking of your brand. You can subscribe to an RSS feed of the search results, or if you want to receive responses as they arrive in your inbox, you may prefer to sign up for TweetBeep (*http://www.tweetbeep.com*), which works very much like Google Alerts.

Technorati (http://technorati.com)

Technorati (Figure 3-3) is a blog search engine that lets you find out what people are saying about you within the blogosphere. The results are available in RSS format, so you can subscribe to the various search results as well.

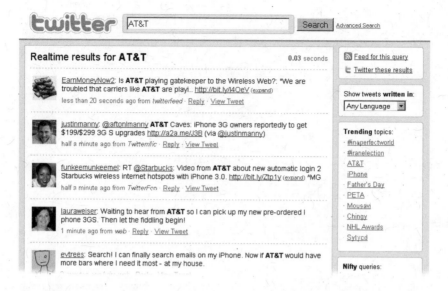

FIGURE 3-2. Twitter's official search engine

FIGURE 3-3. Technorati blog search

Paid Tools

Trackur (http://www.trackur.com)

Trackur (Figure 3-4) is a comprehensive product that costs between $18 and $197 per month, depending on the level of access you request. Trackur provides you with comprehensive social media monitoring on dozens of different social media channels in a single package by giving you a list of all sources in an easy-to-read format. You can view the article, add the source to your favorites, share the article with others, or view (and

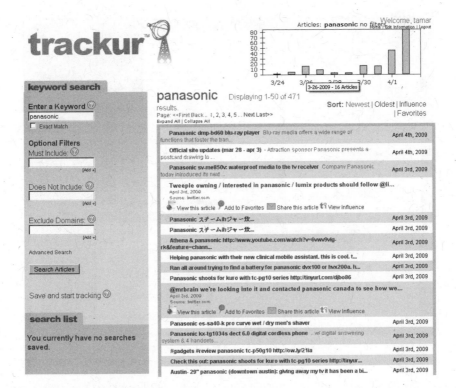

FIGURE 3-4. The Trackur social media monitoring tool

sort) the level of influence of the source, getting more information on how many links are pointing to the particular blog or source and how much traffic the site generates. Armed with this information, you can observe trends and subscribe to results via email and RSS. For offline browsing, the results can then be exported to a CSV file.

Radian6 (http://www.radian6.com)

One of the most powerful corporate tools for social media monitoring is undoubtedly Radian6 (Figure 3-5), which provides monitoring on millions of different social media marketing channels, from blogs to forums to social networking sites. Radian6 users can get in-depth information on social mentions for their products or services across numerous websites; you can also assess the impact of the message through powerful analytics, graphs, and trend analyses. For example, if someone with no followers on Twitter expresses disgust with a product, it won't be as powerful a message as one posted by an individual with 8,000 Twitter followers. When a person with 50,000 blog subscribers endorses a product, it can be a lot more powerful than when a person with 3 subscribers makes the same endorsement.

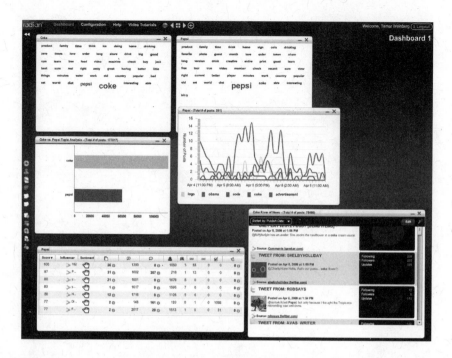

FIGURE 3-5. Radian6 provides a fully customizable dashboard view of your most important social media metrics

Radian6 offers interesting trending data and analytics, and can even help compare the buzz and frequency around different search terms on its fully customizable dashboard.

Radian6 uses a number of different analysis widgets that display on individually customized dashboards. Its River of News widget features a stream of blog posts and comments, tweets, video and photo posts, and other posts across the social web that include your desired search term. You can also access your Conversation Cloud, which is a list of often affiliated with your search, visually weighted in terms of frequency of mention. The Influence widget helps gauge the authority, reach, and social capital of a particular source to help users determine which conversations are having the greatest impact online. Finally, the Topic Trends and Topic Analysis widgets allow you to drill down into keywords and topics, segmenting data by media type, language, or sentiment, and displaying the number of relevant posts by metrics like comment count, inbound links, number of views, or Twitter followers.

Radian6 does not necessarily fit for the small business; pricing structures are scaled based on volume of post results per month and the number of individual user accounts. The service, however, is ideal for agencies, mid- to large-size corporations, and public relations firms.

Socialradar (http://www.socialradar.net)

A premium tool for social media awareness, Socialradar gives individuals monitoring their brands in-depth details and statistics on the fly. Socialradar stores content from 1.5 billion entries on the Web (five times more than Technorati). With Socialradar, you can find out how frequently a specific topic is mentioned, then compare this with other competitive brands in the Trends tool (Figure 3-6).

Another great feature of Socialradar is that it allows you to gauge sentiment about a particular brand. Is the product awesome or does it suck? What kind of terminology is used in conjunction with positive sentiment, and what kind of terminology is being used with the negative sentiment? With the Sentiment tool (Figure 3-7), Socialradar can distinguish between negative and positive posts.

You're also able to view the top sources of any particular product, service, or brand mention and see how they link to one another (Figure 3-8). In this way, you can view the ecosystem of a specific topic and understand exactly which blogs and sources are talking about the topic and how they are related to each other.

Socialradar allows subscribers to click on a source or spike within a graph (for example, any point on the line graph in Figure 3-8) to see relevant posts on the subject matter. You can set up alerts, focus on key sites, build customizable reports, and more.

Among the aforementioned tools, Socialradar is one of the most expensive options, with a cost of a few thousand dollars per month, but it is a great solution for those monitoring big brands.

Other Channels

There are countless other social channels that you can use to monitor the conversation. You may want to follow forum discussions or observe individuals' tagging behavior. Here are some other tools and sites that you should pay attention to:

Video sites

YouTube (*http://www.youtube.com*) and Google Video (*http://www.video.google.com*) let you view user-generated video content.

Photo-sharing sites

Flickr (*http://www.flickr.com*) allows you to see photography and images that have been contributed by users throughout the world.

Blog comments and conversations

Backtype (*http://www.backtype.com*) and BlogPulse (*http://www.blogpulse.com/conversation*) both enable you to find comments and follow the ongoing conversation.

Forums and message boards

BoardTracker (*http://www.boardtracker.com*), BoardReader (*http://www.boardreader.com*), and Omgili (*http://www.omgili.com*) monitor ongoing forum discussions.

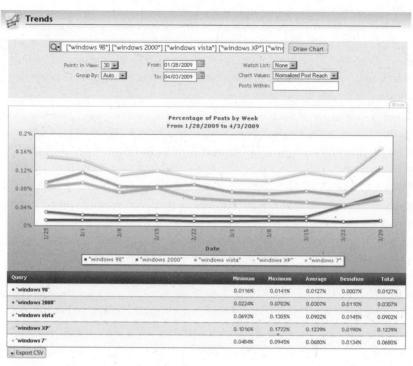

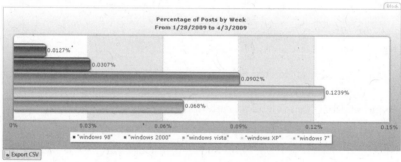

FIGURE 3-6. Socialradar's Trends tool shows the percentage of posts that mention a specific term and displays the output in a line graph and bar graph

Wikipedia

With Wikipedia (*http://en.wikipedia.org*), you can "watch" individual pages and subscribe to the RSS feed of the changes that are made to the page. (Wikipedia is discussed in Chapter 8.)

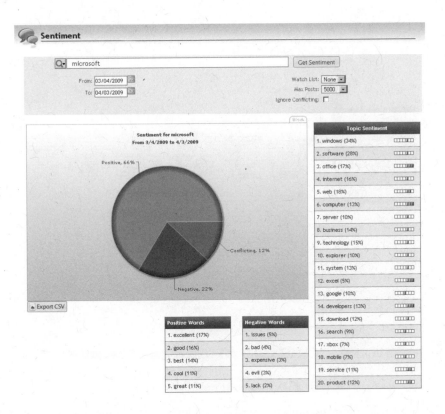

FIGURE 3-7. Socialradar gives you a pie chart of positive and negative sentiment about a brand or product and shows you corresponding terms

Tagging

Keotag (*http://www.keotag.com*) searches tags on Technorati, Blinklist, delicious, Twitter, Google, IceRocket, Blogdigger, TailRank, Live Search, Bluedot, Newsvine, BlogPulse, Bloglines, Digg, reddit, Yahoo!, and YouTube.

You're Listening—What Now?

You're following the conversation, but do those looking to get your attention know this? If you are listening, it's time to respond to show that you are. Whether that means starting your own blog and writing a blog post, creating a video to respond to public sentiment, or having representatives of your company respond to commentary in blog comments themselves, silence is not golden and the public is aching for an official response (in fact, responding with silence actually says a lot!). Let your audience know that you care by taking a proactive stance toward addressing their concerns.

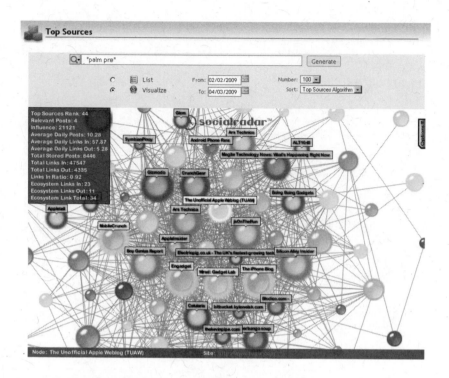

FIGURE 3-8. You can view the top sources of a particular search term with Socialradar

When you do make these attempts to reach out to bloggers and forum members about their feedback, make sure your responses are meaningful and genuine.

Make Sure You're Being Real!

Last fall, Motrin's marketing team created a campaign that ended up inadvertently upsetting young mothers. In the campaign, Motrin claimed its medicine was ideal for mothers who were "wearing [their] babies" and if they "look tired and crazy, people will understand why." The "Motrin Moms" campaign backfired completely and hundreds of mothers expressed their dissatisfaction about the brand through video and blog posts.[2] After several days of silence, Motrin apologized on its website, but that response drew heavy criticism for not being genuine. As influential marketer Seth Godin writes, "This isn't an honest note from a real person. It's the carefully crafted non-statement of a committee. What an opportunity to get personal and connected and build bridges [wasted]."[3] Later, Godin explained that Motrin should have used this experience to be honest and to show that Motrin actually did feel the mothers' pain. Unfortunately, it didn't do that.

In response to how Motrin executed and then stayed silent, microblogging expert Laura Fitton explains the concept of listening[4] as it relates to brand engagement on Twitter:

Even if your brand or agency isn't ready to engage formally and integrate the business applications of Twitter throughout your campaigns, community building and other market engagement efforts, you need to get clued in—fast—to the reasons, times and ways that you can listen. Maybe you're not even ready for full-time social media monitoring. That's your call. But **not tuning in while you launch a new tactic borders on gross negligence**, in this day and age.

This statement applies to any social media outreach attempt. Tune in to the conversations of your consumers on Twitter or on any other social site. If you don't, you're in for a huge disappointment.

Planning Your Social Media Strategy

You're listening and perhaps you're engaging in minor to moderate communication at this point. Now let's consider how you will plan your strategy for execution.

Review the list of goals that you established in the previous chapter. You should have already researched your prospective audience; the sites it visits; the sentiment it has toward your business, brand, and competitors; the tools it frequently uses; and the content it wants to consume online. These questions, which are part of *goal setting*, also relate to the *research* you need to consider for your social media marketing plan. Make sure you have solid answers for each and every one of these questions in order to proceed.

Determine Your Desired End Result

Once you understand your goals and you have performed the necessary research, you have to ask yourself what your desired end result is. We've touched upon possible outcomes of a successful social media strategy in Chapter 2. These include better search rankings, increased brand awareness, and established thought leadership. Your strategy goals might be more tangible ("I want the #1 ranking in Yahoo!" or "I want 3,500 links pointing to my article") as well. If your goals are this tangible, it's important to acknowledge that broader goals may be better: social media is a gamble and you don't know how your audience will respond. I think we'd all agree that Motrin did not expect the mothers to react so negatively to its marketing campaign.

Determine Your Resource Needs

Any marketing strategy will also require blood and manpower. You will need to decide whether to utilize internal resources for the task or to outsource your social media efforts. You may choose to merge the two: you can hire someone in an advisory role over your internal outreach efforts. You may also want to leverage unpaid labor, such as interns, to engage your company in an effective social media strategy. While the younger generation is typically more social media–savvy (and considers this work fun and appealing—especially for free or for college credit), there may be some training required to align your hires with your company's culture.

Any missteps may end up costing your company. Thus, in any social media outreach attempt, training is important, both for your employees and for any external resources you choose.

Brainstorm with Your Team

When considering your strategy, think about teamwork and idea generation. You can strengthen any strategy by utilizing more than one individual for brainstorming. Seattle-based Internet Marketing firm SEOmoz acknowledges that before launching any social media marketing strategy, it has a one-hour brainstorming session with approximately six members of its staff. In these sessions, any ideas related to marketing the product are discussed, no matter how crazy they may sound. No silly ideas are thrown out, because after all, that content may be just what the target audience is looking for.

Mind mapping

Mind mapping is a terrific approach for brainstorming. The idea behind it is to visualize your ideas by placing a single thought in the center of a page and then branching out to related themes through word association. An example mind map, taken from an individual's Flickr stream,[5] is shown in Figure 3-9.

FIGURE 3-9. A mind map

Navigate the red tape

Finally, you may have to cut through corporate red tape in order to plan your strategy. Perhaps you'll have to run your social media strategy attempt by your legal department. Is it aware of the pros and cons of social media engagement? You may also need to get through multiple levels of bureaucracy, and several key individuals may be required to sign off on any strategy.

The problem with corporate red tape is that the hoops you have to go through may hinder progress and ultimately may not work in your favor. Whether or not your legal team gives you the green light, your brand will be discussed. It is important to focus on training the individuals within your company so that they represent your company's missions and goals in a factual light with honesty and transparency; the application of these training methods will be explored throughout this book. Training your corporate team may also be a necessary part of this equation.

Networking Within a Social Medium

Once you have planned your strategy, you will need to execute it. You must acknowledge that you will be relating to individuals, and you will be able to measure success by determining how other people react to your outreach attempt. Your strategy will almost always include promoting your agenda alongside a community, and as such, you must consider networking.

Social media is not that different from real relationships. In both, you are dealing with real people. If you get off on the wrong foot in your social media relationships, they may not last, just as a personal real-life relationship can fail with the wrong first impression (or if you commit a subsequent egregious mistake). Further, since many savvy Internet users do not easily trust outside influence, one single online error can destroy your relationship, especially because it is much harder to redeem yourself from a blunder when body language is absent and genuine apologies may not necessarily be perceived as genuine. While we are sitting behind screens, we may be losing sight of the fact that the recipient of our personal messages has emotions and feelings. In a social media relationship, you need to be authentic just as if you were in a face-to-face relationship. You never know how or if these relationships will end; if you get involved enough in online communities, they may materialize into real-life relationships. Hundreds of thousands of face-to-face meetings have arisen as a result of social media and the communities that unite like-minded individuals within.

Networking both online and offline can be very rewarding. In fact, networking is probably the most important part of the social media equation. The key to networking efficiently is to understand the communities you market to and to give of yourself, at which point you can establish yourself as a community manager who represents your brand or, on a smaller scale, you can build a *power account*—that is, a strong respected account on a social network—that can be used again and again.

Understanding and Listening to Your Audience

With social media marketing, knowing the types of people you are going to interact with is of extreme significance. Understanding your community is critical to achieving your goals.

The Newcomer's Parable

Imagine you're an immigrant in a foreign country and you want to fulfill the "dream." However, you don't know anyone in this new land and you are afraid of taking the next step. Fortunately for you, you have similar interests as the inhabitants and can share them with the community when the time presents itself. As you begin your journey, you'll run into obstacles and roadblocks, but as long as you are determined and opportunity knocks, you share among your peers. You have skills that your neighbors appreciate and they, in turn, help you out as well. Over time, you'll grow as a member of the tribe and will no longer be viewed as the "new kid on the block." Perhaps, in due time, you'll even become a shining star, a leader, or a notable member of the community.

The previous paragraph is from a blog post I wrote articulating how social media is truly about community.[6] There are millions of social media consumers who visit hundreds of thousands of websites. These consumers are choosing sites to frequent based on the members of the communities therein. In social media marketing, or in any developing relationship, it's important to understand your surroundings, especially if you're a newcomer. Acquaint yourself with the community and really understand it. You need to:

- Know the demographic of the individuals you are pitching to
- Know the community's likes
- Know the community's dislikes
- Contribute on a consistent basis to the community constituents
- Establish yourself as an active participant of the community

If the users within the community, for example, abhor those who represent themselves as "marketers," you may want to interact with its constituents as if you are associated with your company, but not in a marketing capacity. After all, if you do identify yourself as a marketer or convey your intentions as, "I'm selling this product," or, "I want you all to know that I have a great tool coming out," your overt selling tactics will fall on deaf ears and may even get you ostracized. As this chapter illustrates, you should become a community participant and not a salesperson. The culture has changed. The landscape is now that of goodwill without overt commercial overtones. In fact, all communities begin without marketers. When marketers later realize that they can capitalize on the "unsuspecting" individuals, the problem of trust develops as communities blossom with newcomers looking to make a quick buck. If your strategy is to infiltrate the community solely to spread your marketing message, you'll make no friends and cause the community constituents to distrust the practice of marketing even further, perhaps dubbing it a "snake oil" practice, a view that is already rampant on the Internet.

Help Your Friends

Communities flourish because people are helping other people. If you want to be an accepted new member of the community, you need to think about the wants and needs of each individual user and the community as a collective before pushing your personal agenda. Later, your "ulterior motive" can be communicated (just as long as you continue giving back to the community and its members look up to you as a respected contributor), but it's more important to establish yourself as a reputable member who wants to give back to the community first. Once you do, you can begin to take, as long as the community is receptive and wants to know more about you as a community participant, but you should always keep giving.

Yes, this practice takes time. However, as far as networking is concerned, this is time that you should be willing to spend to better build your brand and to keep your reputation as an individual looking to market a specific product on a clean slate. Leveraging these connections can help you position yourself as an expert in the particular subject matter you intend to promote as well. Continuing to interact with community members can also foster real-life relationships.

As this book has touched upon in the previous chapters, social media also provides a way to combat a reputation management fiasco. If you are part of the community and build relationships when you don't need to, you will have more credibility when you really need support.

Social media communities are real relationships, real conversations, and as such, they should be treated like they are real. It's not about a "me, myself, and I" mentality. It's about the collective, the community, and the common good.

As my friend Anna once said, "These real conversations can lead to *real* business, so pay attention."

The Importance of Giving

Too many marketers approach social media as a goal-achieving tool. You have your goals from Chapter 2 outlined, right? If so, you might look over your list and realize that you just want to get your social media campaign over with and cut corners to achieve those objectives. If this is the tactic you are taking, you may want to elect someone who will enjoy his/her engagement in the social media space. This does not have to be a chore, nor should it be perceived as such. Too many people consider social media marketing as *only* a way to get more eyeballs, links, or awareness. In reality, social media is about those real relationships, and it should be fun.

In the previous section, you learned about the users who will be consuming your social media content. In any community, participants will receive outreach attempts in either a positive or negative manner. There are those who will embrace your message and others who will be skeptical of any outreach attempts at all. In all of your attempts, though, your goals should be about *providing to someone else* rather than benefiting yourself.

Ask yourself not about what you will *get* from social media marketing, but what you can *give* of yourself and your expertise. Rob Key, CEO of social media marketing and communications agency Converseon, says that you should "lead with altruism" and "come bearing gifts." How can you utilize these social media channels to give back to the community participants? How can you help members of the community with problems they may face? If you think about the community *first*, you will ultimately be able to achieve all of your original objectives in a way that will allow you to earn trust and respect from those you are targeting with your messages.

If you have not put your constituents on a pedestal, do so now. Being viewed in a positive light on a regular basis will only happen if you consistently value your community, listen, and give back. In social media, this is the kind of behavior that can put you miles ahead of your competitors.

Community Managers

If you're going to be consistently pushing social media strategy within your company, you may want to enlist someone in a new and emerging role that is becoming a lot more necessary within the social sphere. The *community manager*, an individual who works alongside the marketing team (or perhaps is a member of the marketing team), is often the public voice of the company and also connects with customers and shareholders on a personal level. At the same time, the community manager is responsible for conveying the feelings of the customers to internal departments.

The community manager should be an enthusiastic team player who can develop a sense of awareness about your particular brand within the industry and related social circles. The more important part of this is that the community manager doesn't perform like a regular marketer: she is actually responsible for making outreach efforts that do not sound like they're coming from a marketing department. Instead, these efforts align with the community's interests without commercial overtones. The community manager takes a communications role that is about fostering solid relationships. Her role is multifaceted; the ideal community manager should be able to communicate both in person and online, and should be someone who loves working with people, is personable, approachable, has a sense of humor, and can enjoy the challenges presented to her. After all, the community manager has the key responsibility of humanizing the company.

Community managers use social media very frequently for monitoring and maintaining relationships, and thus ideally they should be regular bloggers/videographers on behalf of the company. They should also maintain a presence on a variety of social media sites. The community manager monitors the conversation and participates in it on a regular basis. Using feedback given by constituents, she analyzes anomalies, patterns, and trends that may be of concern and escalates them to the company's executive team.

At the same time, community managers may want to identify the advocates of the product or brand. Who talks about the company in the most positive light? Who doesn't? What can we do to "convert" those individuals over to brand evangelists? Those individuals can then be offered more perks for working alongside the company.

THE BRAND EVANGELIST

Brand evangelists, or brand ambassadors, take your product seriously. They are heavy users of your product and care a great deal about it. They want your brand to succeed. That's why they can be spotted in the wild: they talk, and you notice. Because they are part of your target audience, they probably know your target audience better than you do. As such, they are users that you'd be foolish to ignore.

To find those brand evangelists, you first need to determine who they are. Perhaps they're blogging about your product or they're very vocal in the conversation and show a passion for your product by uploading videos of themselves using it to YouTube. Reach out to them. Once the individuals have been selected, find out what they love about your brand and what is lacking in your product development. Take their feedback seriously—even their criticisms—so that you can create better products. Engage them by giving them special access and privileges, and empower them with communication channels. You can also reward brand evangelists by featuring them as content creators. Even small gestures are appreciated. One company, Startup Schwag, features photos of its products on its main website as long as they are uploaded to Flickr and tagged as "startupschwag."

How do you keep your brand ambassadors happy? Be accessible and reachable. By giving them the ability to contact you easily, you can even discover other potential brand evangelists. Also, it's great to create communities for the ambassadors both online and offline.

Community managers are essentially social media marketers of the organization. They operate best following as many communities as possible, maintaining social media profiles across a variety of networks, maintaining a microblogging (Twitter) account that has real value, and commenting on blogs. The model community manager subscribes to alerts about the brand (for a variety of key phrases including competitors, industry-specific keywords, and the brand/product names), reads relevant blogs and news outlets daily, follows related individuals in the space (including prospective and current brand evangelists), and if applicable, follows related podcasts and video channels.

While being a regular participant is crucial, the community manager does not outright identify herself as such. Instead, she takes a vested interest in the topic at hand and offers value in any volunteered commentary. Blogging consistently on behalf of the company and product comes next. For more on blogging, see Chapter 5.

The community manager is required to be a proactive voice on behalf of the company at all times. An ideal community manager should respond to substantial concerns immediately (within 24 hours). In social networks, momentum dies down when visibility dies down, so responding to user concerns immediately, as in the Motrin Moms scenario, is important. In a week's time, the attention likely has shifted elsewhere.

Over time, the community manager should effectively articulate the mission of the company. She should raise awareness through consistent blogging. It is also in her best interest to highlight appropriate user comments and concerns and follow up in a blog post. The responsibility of community management does not necessarily need to lie in the hands of a single individual. All individuals in the company can have a role in communication. At the minimum, one individual should have a greater role in monitoring, participating, and analyzing, but several companies have embraced the notion of the "employee evangelist," which translates to many happy employees and customers.

Power Accounts

Community managers have a huge role in communicating with community constituents and internally within the organization. Yet there are very few community managers who can consider themselves "social media power users" in the sense that they are extremely avid users of social media sites. Many community managers only participate in these communities by responding to company-specific concerns at the maximum, but there exists a group of hardcore social media users who have gained nearly complete trust on a variety of social media sites by being extremely active. These individuals are users who maintain *power accounts*. Power account holders are mostly dominant on social news sites where contributions are highlighted on the front page, or through profile pages that show that they are "top users" based on the number of successful submissions they have achieved over a period of time. In fact, on popular social news site Digg.com, the top 100 most influential users control between 43% and 56% of the home page content.[7] Establishing yourself as a top user can mean tremendous opportunities for success.

Building a power account is optional and does not necessarily need to be the responsibility of the community manager. In fact, individuals who hold this kind of power often get solicited for additional consulting and promotional work, so it may be helpful to outsource this role entirely or to deal internally with the possibility that "with great power comes great responsibility." If company policy prohibits it, you should ensure that the individual being held responsible to maintain the power account should not also use that account for personal gain on the side.

Power users are often considered "hardcore." Essentially, they are similar to brand evangelists of the social sites upon which they have already achieved greatness. A community manager in such a role clearly should understand the issues she is facing by being an evangelist of another community.

How do you build a power account? The simplest answer is through very regular and very consistent activity. There's a lot more to this, however. The most influential power account holders have established themselves as very active contributors of social news sites. They do this through networking, making quality submissions to the website, and consistent engagement.

There are two major considerations when building up a power account.

- Who should maintain the account? Should it be owned by a company or by an individual? Ask yourself whether the company should own this social media profile or if you should let the current individual working on behalf of the company maintain this account. It may be extremely valuable in the long haul for the company to own the account, but it may only be influential when manned by the particular individual who has already fostered its success.

- Should the account be associated with the company or brand? No, you should not identify this power account with a particular company and a brand. Those types of accounts appear to be created only for the purpose of marketing your product. They will not be easily trusted if they are associated with a particular marketing message.

Ten Commandments of Power Account Submitters

Power account holders live by a set of rules that help them build credibility, establish identity, and make them memorable among the community. Anyone, however, can follow these rules on social news sites to become a respected and valued participant of the service.

#1: Thou shalt distinguish thyself with an avatar

Submitters on a social news site should immediately upload an avatar if they want to be considered legitimate by the established authority (that is, other power users). Avatars do more than just make you stand out in the crowd: they help to establish your identity, which is vital if you want to be a memorable contributor of a social news site. Don't just blend into the background; power users assert authority implicitly by establishing identities that help others remember them. Focus on louder colors that do not necessarily blend in with the background of the site or appear similar to the site's default avatar.

> ### DEFINITION
> An *avatar* is a small image that is typically displayed next to a username to distinguish users on social media sites and forums.

#2: Thou shalt be genuine

Power users need to be perceived as altruistic submitters. You should not consistently promote your own content. If you want others to trust you, you must consider the content the community will like. Be a real resource and not a self-serving one. If you consistently push content in a particular subject matter (for example, you're passionate about politics and you end up sharing interesting content among the thousands of users on a particular social site), you will become an established authority on the subject matter and people will follow you. However, you might feel more comfortable if you diversify your submissions and use multiple high-quality sources.

#3: Thou shalt network

Of critical importance in establishing a power account is networking. Many people would argue today that social news sites are only "social" sites, and not "news" sites; it's all about the relationships you have forged over the course of your involvement on the site. You may submit a great news story to a social news site, but if nobody else sees that story (and nobody else is alerted to its existence), there is no chance that the story will hit the front page. If nobody knows you, then your submission, no matter how powerful it is, will fall into oblivion.

The ideal solution would be to network with those who have already established themselves as the power players. If a username is frequently visible on the front page of a social news site, chances are he or she is a dedicated user. If you are truly interested in becoming a dedicated user, you should take the initiative to introduce yourself and get to know that individual. It's important, though, not to immediately let it be known that you are here to solicit votes from users who have already built trust and credibility. As a power user on numerous social sites myself, I've been contacted more than once with "hi, I see you're active on Digg! Can you please Digg my story?" That's *not* the right way to network, under any circumstance.

If you're looking to build up a credible power account, the users you're aiming to connect with are going to be your friends for the long haul. Networking is clearly not an overnight process, and you should be diligent in how you approach the power users who everyone wants to become friends with.

#4: Thou shalt submit high-quality stories to the social sites

As you know, "high quality" is subjective. What if the story I spent 30 hours writing is not deemed appropriate by the community? The important thing is to study the community first and understand what it likes. High quality often has some criteria, however: submitting well-written content is critical. Don't submit irrelevant spam or short blurbs. If you can find the original source (the source that broke the story and that other sources subsequently reference), by all means, submit that instead (if it wasn't submitted already).

#5: Thou shalt be fast

Real power users have a good idea when stories are published on the original sites. Some popular news outlets have an editorial schedule: features get published at 12 noon or 12 midnight. The most avid of social news users will know as soon as a story is published, and will submit it to the social news site immediately. There's no time to wait because there's huge competition to get the story submitted first.

When news breaks (for example, a terror attack in a blossoming city, the passing of a great actor, the winner of the World Series, or the announcement of a new Apple product), be the first one to submit the story as long as it's from a credible source. The news story may still be developing on the original source, but that means that there will be good content on the news submission in due time.

Of course, you can't always tell when some of those news stories will break, since breaking news is typically unexpected. Fortunately, Apple often releases products during Tuesday keynotes. Popular technology sites like Engadget and Gizmodo cover the product releases in real time.

#6: Thou shalt study the sources that have achieved greatness on social sites to understand what the community likes

How do you determine what is "high quality" on social news sites? You study what the community prefers to submit. Check the popular domains. This may require your own research. However, an independently run website (*http://www.di66.net*) shows Diggs' top domains, top titles, and top descriptions according to front-page success.[8]

#7: Thou shalt dedicate time to the task

The key to being a power user is to spend a considerable amount of time on the social sites. This allows you to validate your intentions to the community at large. A typical power user initially spends approximately three to four hours a day building credibility. Once you gain trustworthiness, you will still need to maintain a time commitment, but not as substantially. Some power users who have established themselves as such check their accounts only once in the morning and once in the evening.

#8: Thou shalt help thine friends

Regardless of whether the rules dictate that you shouldn't always vote on your friends' stories, backscratching is very prevalent on social sites. It's all about building those relationships. If you consistently push your friends' content, they'll be a lot more likely to help you out, too. Again, these sites are *social* in nature.

Pay attention to individuals who are voting up your stories long before they ever hit the front page, because those are individuals who trust you but likely also want to be noticed. Establish solid relationships with these users.

As a precautionary note, however, be careful about too much reciprocation; sometimes, complete mutual support can lessen the impact of success.

#9: Thou shalt use consistent account names over all social networks

Once you've achieved greatness on one social network, users will follow you on other social networks. This increases the opportunities you will have for networking in the future and will possibly give you a boost in these other social networks as an established user. Of course, you will still need to study the community mentality and make sure you're following the norms, but this becomes a lot easier if you are recognized across several social networks.

#10: Thou shalt use other social networks for inspiration

Cross-pollination—that is, submitting stories across several social networks—is a great way to achieve success on social sites. While not every reddit front-page story will be successful on Digg (since the interests of the communities differ), there is a good chance that those stories will be noticed. Plus, if you can spot a great story that is growing in momentum on one social site, it could still be well received on the other social site. Finally, looking on other social sites for great submission ideas is often easier than competing against the 30 or so users who are all anxiously trying to be the first one to submit that Lifehacker noon feature.

Should I Become a Power User?

It's easy to say that power users have really achieved success over time given their involvement on social news sites. They have the audience; individuals follow their submissions closely. They have a good sense of the stories that succeed on social sites. They are trusted contributors.

But now that you have reviewed those 10 commandments, it should be apparent that it's a lot of work! In fact, it's so time-consuming that power accounts do not exist as much as they did in past years, and much of this relates to algorithmic changes that diminish the impact of these account holders.

At a recent conference I attended, successful blogger Brian Clark wrote that "you can either become a power account user or know someone who already is one." Clearly, some of the most influential marketers have networked with the power users and have achieved awareness and greatness. They did not become power users themselves. With the number of influencers who have contributed thousands of hours to the task, it's evident that there are already users who you can turn to. This approach is certainly a lot easier than building up your power account from square one.

Of course, if you don't have loyalties to a specific company, or if you just want to establish yourself among several social communities, a power account is still a viable choice. More often than not, however, community managers are a more realistic choice for longer-term growth and success, especially as they are prevalent on multiple social channels and do not necessarily have to focus their energies on social news sites exclusively.

Summary

When executing a successful social media strategy, listen to your constituents across as many social networks as possible. Build relationships with individuals on social sites and establish a brand identity in those parts. Above all, listen to the community and act upon its feedback to create better products that users are actively seeking.

Since people can be talking about you nearly everywhere, it is hard to keep track of what they are saying and where they are saying it. There are numerous tools that you can use to monitor the conversation.

A community manager may wish to establish communications between the company and the customers. She loves working with people, loves a challenge, and is an effective communicator. The community manager may also wish to become an influential power player on social news sites by actively engaging on them. Once she gains the trust of the community, she can slowly push the agenda of the company by submitting relevant stories that will be well received by the community.

Power account holders are most visible on social news sites since their immense dedication has brought them the respect of other community members. To become a power account holder, there are a few guidelines you can follow, though the time and energy commitment is substantial. Even if maintaining a power account is not a strategy you are interested in, understanding the 10 commandments of power account submitters can help you succeed on these social sites.

Endnotes

1. *http://www.israelpolitik.org/about/*

2. *http://www.parenting.blogs.nytimes.com/2008/11/17/moms-and-motrin*

3. *http://www.sethgodin.typepad.com/seths_blog/2008/11/we-feel-your-pa.html*

4. *http://www.pistachioconsulting.com/motrins-twitter-moment*

5. *http://www.flickr.com/photos/claypole/2074087698*

6. *http://www.techipedia.com/2007/community-participants-rock-my-socks*

7. *www.pronetadvertising.com/articles/the-power-of-digg-top-users-one-year-later34409.html* and *www.seomoz.org/blog/top-100-digg-users-control-56-of-diggs-homepage-content*

8. *http://www.di66.net/top-sites-365d-by-posts.html*

CHAPTER FOUR

Participation Is Marketing: Getting into the Game

Participation is essential for success within social media marketing campaigns. As you've learned in the previous chapters, constant engagement is necessary, but real conversations are even more important. The most effective form of social media marketing requires relationship building in a completely authentic way. In this chapter, we review several case studies and show how companies can empower themselves by engaging in social media marketing. We also look at how you can avert or more successfully address reputation management issues via social media marketing channels.

The Cluetrain Manifesto: Markets Are Conversations

In April 1999, several marketing masterminds released a precursor to the social media marketing of today in *The Cluetrain Manifesto* (Basic Books), a manuscript of 95 marketing theses. The message of *The Cluetrain Manifesto* is simple but very powerful: markets converse with one another and the Internet has facilitated this communication. Because the Internet has a profound impact on both consumers and organizations, with improved communication across Internet channels comes the responsibility for organizations to adapt to an evolving environment.

TIP

The 95 theses of *The Cluetrain Manifesto* are available for free online at *www.cluetrain.com*. Additionally, a print copy of the book is available at bookstores.

Given that social media marketing as a practice is still an emerging technology, *The Cluetrain Manifesto* was years ahead of its time. We're now beyond the 10-year anniversary of its publication and its message is as important as ever. While the authors of *The Cluetrain Manifesto* were able to accurately predict a marketing revolution, only a small fraction of individuals have actually jumped on the bandwagon. Social media marketing and Internet-facilitated conversations are still not fully mainstream and the importance of Cluetrain's 1999 message is not necessarily being emphasized. There's still time, and with 10 years having passed since the document was published, all markets should really understand that this is the prime time to create real, valuable conversations.

If you're the first person in your industry to engage in social media, by all means, don't let it hold you back. You can lead by example.

The "Participation Is Marketing" Phenomenon

Back in 2007, new media marketer Chris Heuer coined the phrase "participation is marketing." Heuer asserts that the best marketing minds are those who participate in the communities they service and don't just aim to sell products directly to the people. After all, companies and organizations exist to help people with a specific problem; the original purpose of these organizations was to participate in order to address a particular need. Moreover, pitching products and services is an outdated tactic that will not be well received among individuals who have either grown tired of the same old marketing message strategies or who have gotten accustomed to the newer tactics of social media engagement.

I'm going to take the phenomenon a step further and argue that the phenomenon of "participation is marketing" also works the other way. While marketing folk may be participating in communities—and that translates to "participation is marketing" (especially as the involvement humanizes the company and is not as self-serving if done appropriately)—I'd argue that community members who are using social media to participate directly with these marketing people are also supporting the claim of "participation is marketing" because they are showing an interest in being part of the greater marketing message. Participation, again, goes both ways. If marketers engage in a community that is not frequented by consumers, they are not serving the greater community and the consumers are not participating in the marketing message. Of course, to realize the idea of "participation is marketing," the consumer must actually engage in the marketing message, whether by commenting, by vote endorsements, or by referrals, but two-way communication is crucial. In this regard, loyalty is built by fostering genuine relationships with the members of the community. Chris Heuer says:

> If you are trying to sell something to the community, and that is your reason for being there, it will be obvious to those people and you will never be as successful as you can be. If you are participating because you really want to contribute to the community, because you really want to share what you know, because you really want to be of service to the community and its members, you will sell to the right people BECAUSE of your sincerity and honesty.

Today, trust is not easily bought with advertising dollars. Not anymore. Establishing relationships is a critical element to the notion that "participation is marketing." In the previous chapters, we looked at the importance of monitoring the conversation and responding to feedback. The key here is to actually follow communities where your brand, product, and service are mentioned, and to engage altruistically with real interpersonal human interactions. It's not enough to maintain a blog or to respond by using the Press section of your website to "participate" (this was part of the tremendous misstep in the Motrin Moms campaign that was discussed in Chapter 3). Participation requires ongoing and real dialogue on a consistent basis. If you're ready to engage (and hopefully by reading this far, you've made the decision to participate), you should consider a comprehensive strategy that accommodates users in communities where your products are discussed. You also should be completely genuine at all times. Further, if you immediately spot a perception issue about your brand, don't immediately push your message onto the community constituents. Walk a mile in their shoes first to understand their mindset and mentality. Later in this chapter, we'll explore case studies illustrating the "participation is marketing" phenomenon.

Participation Is Marketing for Public Relations Professionals

The area of public relations is also seeing a tremendous evolution and paradigm shift. Communicating on behalf of a client to a group of constituents via snail mail, phone, or email is obsolete. In keeping with the "participation is marketing" theme, public relations professionals should actually have relationships with the public, which goes beyond the standard press release. Further, online communities are becoming more influential than traditional media outlets. A successful social media campaign can immediately raise product awareness for hundreds of thousands of individuals who are specifically seeking out these communities for relevant news. On the other hand, these same social media consumers are not looking for the traditional pitch. They are looking for newsworthy content that the community as a whole believes is relevant to a wider audience. They rely on the valuable and selfless contributions made by reputable members of these sites and communities.

Public relations professionals will need to up the ante in terms of how they communicate on behalf of their clients. The ideal PR professional is a community participant and not just someone who is solicited to send out an often irrelevant pitch to a publication.

Graco Baby Case Study: "Participation Is Marketing" Translates to Brand Awareness and Exposure

In 2007, social media agency Converseon had the challenge of creating a social media strategy for 66-year-old brand Graco Children's Products. The challenge: to determine how an old brand could interact with young parents who use the Internet. How could Graco find individuals to target? Most significantly, how could it tap into these communities without appearing overtly commercial?

The first thing that Graco realized was that it needed to research the social media space. In time, Graco discovered that there are numerous parents using message boards and forums, that there are a multitude of Mommy Bloggers, and that these parents are also engaged on other social networks, such as Twitter and YouTube.

To gain awareness, Graco determined that it should establish long-term relationships with active parents engaged in these online communities. Naturally, the goal was for Graco to make its way into these communities authentically. After discovering the online parenting communities, Graco listened to the ongoing conversations within these networks. Doing this helped it devise a strategy that made it an accepted participant in the ongoing conversation and humanized the entire Graco brand. After all, many of Graco's employees are also parents looking out for the best interests of their children.

By listening (or "conversation mining"), Graco discovered that the parenting communities were resistant to commercial endeavors. Thus, it learned that altruism was important; instead of forcing its brand name into the communities, the company and Converseon ended up using Graco's name in tandem with real-life events and get-togethers for parents. This specific environment was more accepting of the Graco brand name, and it became a great way to capture the minds of influencers in the parenting space. It was also an appropriate avenue for Graco to engage with the community. After these in-person meetings, a parent-centric blog was launched. Multiple Graco employees currently contribute to the blogging effort.

Graco didn't stop there. It also engages in Flickr (*www.flickr.com/photos/gracobaby*; shown in Figure 4-1), Twitter (*http://twitter.com/gracobaby*; shown in Figure 4-2), and YouTube.

FIGURE 4-1. Graco's presence on Flickr: Bringing the community together

FIGURE 4-2. Graco offers advice for and responds to parents on Twitter

To date, Graco has seen a lot of successes with its social media strategies. After monitoring the conversation, Graco observed that the brand was discussed in a more positive light (up to 83% in 2008 from 68% in 2007). All posts regarding its outreach efforts (100%) have been positive. Graco observed a substantial increase positive reviews and recommendations for online products. Graco's social media success has landed it an appearance on the *Today Show* and in several online publications.

Within six months of the Graco Baby blog launch, it was ranked the 59th best parenting blog in the ParentPower Index.[1] In May 2009, the blog was ranked #34. Graco's blog drove traffic and visitors and, of course, yielded increased conversions. Graco's blog enabled it to get a better ranking for competitive search phrases in the search engine results, and the blog is now considered a leading authority according to blog ranking service Technorati.[2]

Today, Graco is truly embracing social media. Lindsay Lebresco, Public Relations and Social Media Manager of Graco, says, "We see social media as a new way to create a stronger, deeper and long-term relationship with our consumers. Social media gives our brand access to our consumers and their needs in a unique environment and allows our messages to be delivered in a personal, transparent and immediate way."[3]

Tyson Foods Case Study: We Have a Blog and We'll Use It for Good

If you associate your brand with something positive, it can result in good feelings toward your company. Tyson Foods took this idea to the next level when it launched its Tyson Foods Hunger Relief blog in late 2007 (*http://hungerrelief.tyson.com*). Ed Nicholson, Director of Community and Public Relations at Tyson Foods, explains that the company, which has been involved in hunger relief since 2000, recognized that there was a passionate community of people and organizations who work selflessly on hunger initiatives on a national and local level, but that there was no social media space that catered to these individuals. Further, awareness of the widespread nature of hunger is not very high. Nicholson says, "We know people who have cancer and heart disease. We can personally and emotionally engage and relate to illness. With hunger, it's not so easy. Once people become more aware and engaged in hunger issues, they have more of an emotional attachment to the subject." The blog was thus launched to raise awareness of hunger among the American population (Figure 4-3).

From the start of the social media initiative, Tyson realized that it needed to be authentically engaged. Its efforts are completely transparent and genuine. In fact, beyond highlighting shocking hunger statistics and sharing tips on hunger strategy, the Tyson Foods Hunger Relief blogs also shares stories of positive hunger outreach attempts throughout the country in a category dedicated to bringing to light the contributions of numerous "Hunger All-Stars" (*http://hungerrelief.tyson.com/AllStars*). For example, the All-Stars page recently discussed the efforts of 13-year old Jonathan Crider, who raised $20,000 for his local Oklahoma City food bank, and told the story of Ethel Shepherd, now 80, who started a food pantry 25 years ago out of the basement of her church and is continuing her mission today.

FIGURE 4-3. Tyson Foods' hunger relief blog

The hunger awareness efforts have not ended with the mere launch of a blog. In a blog post published in August 2008, Tyson Foods invited the community to participate in a food donation of epic proportions. For every comment left on the blog post, the company promised to donate 100 pounds of food to a food bank in Austin, Texas. Within six hours, an entire truck had been filled. There are more than 650 comments on the blog post to date.

In December, Tyson launched a more powerful hunger giveaway to the Greater Boston Food Bank. Once announced, it took less than two hours for the first truck to be filled to the brim with food. A second truck was also filled by the end of the day. There are over 900 comments on the blog post.

Beyond the blog, Tyson is also actively involved in social media. It uses Twitter to communicate its philanthropic efforts (*http://twitter.com/tysonfoods*, shown in Figure 4-4). It also shares its outreach attempts via Flickr (g (*http://www.flickr.com/photos/tysonfoods*)). Nicholson maintains LinkedIn and Facebook accounts that he acknowledges are integrated with his

NewYorkers: Hunger in your hometown. Comment we'll donate 100 lbs. to @FoodBank4NYC http://tinyurl.com/77r56k

4:12 PM Jan 13th from twhirl

TysonFoods
Tyson Foods

FIGURE 4-4. Tyson Foods uses Twitter to promote food donations

activity at Tyson Foods. All of these specific social media accounts have contributed to awareness of the Tyson Foods brand.

Tyson continues to update its blog with posts about its charitable contributions. But Tyson also finds that its positive engagement has amounted to success for the Tyson brand itself. The company has seen increased engagement among the community and improved relationships.

Tyson's outreach efforts are still new, yet Nicholson notes that Tyson will continue to grow the communities to become a leader in this area. Nicholson adds, "We don't give the most money but we have this product we can give." Tyson Foods will continue to leverage the communities it has, to be a part of the community, and to become an important player in the social media space.

The Home Depot Case Study: Tapping into the Mindshare of Valued Customers

Due to its initial fear of the uncontrolled message, The Home Depot took a logical approach toward social media with several slowly evolving initiatives. Based on the success of these programs, the company delved deeper into the social media space. In the end, these initiatives, beginning with product reviews and expanding to social media channels, including YouTube and Twitter, have resulted in a lot of success for the retailer.

Nick Ayres, an Interactive Marketing Manager at The Home Depot, has been at the company for over three years and has been instrumental in helping it seek out several different social media outreach campaigns, which have all shown varying levels of success.

The first step toward social media involvement and awareness for The Home Depot was encouraging the community to write reviews on products marketed and sold by the company. Using this tactic, The Home Depot initially saw how customers involve themselves in the social space. As Ayres notes, The Home Depot was initially fearful of the unknown, especially the negative, so offering product reviews was a preliminary baby step into the world of social media. He says, "If we opened the floodgates of letting customers tell us about the Home Depot brand, we were afraid to see a lot more negative than positive. It's a lot less of a concern when you start with product reviews as that's a logical extension of what we've been doing." With product reviews, users have the ability to evaluate their purchases in the same way others do on a big front-facing site like Amazon.com.

Response to product reviews has been overwhelmingly positive, he says. "This is one of the few areas where we can tactically say that there's a good financial upswing for us that you can really see in the general marketplace. Generally speaking, products that have higher ratings and reviews do better from a sales perspective."

The Home Depot's next social media venture was in leveraging video sites to teach users how to do things in the house. The Home Depot has an official YouTube channel that showcases video demonstrations of do-it-yourself home projects (*www.youtube.com/user/HomeDepot*, shown in Figure 4-5). The video library is relatively robust, with over 80 videos and hundreds of subscribers. Video topics range from home improvement to energy efficiency. As for why YouTube and other video sites were preferred channels for The Home Depot, Ayres notes that the content has always been at customers' disposal, but, "We wanted to engage with a different audience in a much more real-time and personal way."

Additionally, The Home Depot uses Twitter (*http://twitter.com/thehomedepot*) to broadcast company updates and communicate with users, particularly for customer service issues. In the summer of 2008, Twitter was abuzz with news on Hurricane Gustav. The Home Depot capitalized on the opportunity to communicate accurate store information as it related to hurricane preparation. For instance, the company's Twitter account was used to communicate store closings and offer information on the availability of essential supplies in local stores. The Twitter stream was not directly used to sell products, but rather to articulate information and availability to those needing timely updates. Plus, for the retailer, the information was already readily available, as The Home Depot had a well-defined hurricane process at its hurricane headquarters. The hurricane information was available to in-store customers and online on The Home Depot's website, but was later disseminated in bite-size chunks and passed along to the general public using Twitter.

FIGURE 4-5. The Home Depot's YouTube channel page

Ayres finds that The Home Depot is successful because it's using already available information to arm the public with knowledge: "Our biggest wins to date have been not trying to reinvent the wheel, but by taking what we knew worked and trying to find the places where that information and products made the most sense in the social media/Web 2.0 world. We've seen pretty positive reactions because of the way we went about our social media initiatives strategically from the start. We are always taking feedback and our successes to invest in the future."

In the future, The Home Depot is going to extend its "digital orange apron" (its signature in-store "uniform" is the orange apron) to provide consistent customer experience across all media channels and to serve this information up to a whole generation of customers. Armed with "digital orange aprons," the company will be able to take the resident knowledge of employees and associates and present it in a way that will be more accessible to customers engaging in the social media space.

Ayres himself is involved in social media; he mentions that he maintains his own Twitter account, but that The Home Depot has an official channel that is managed by its Corporate Communications team. He acknowledges that before The Home Depot made its debut on

Twitter, the retailer asked the community how the microblogging service should be used, because it was not sure whether to use the Twitter account to offer sales messages or for customer service. The community responded that it would prefer the account to be focused on addressing customer concerns, and The Home Depot listened. Therefore, Ayres may use his own personal Twitter account to circulate The Home Depot messages from time to time. However, he notes that while he will sometimes reach out to customers to assist them with issues, he will always do so in a transparent way. It is important for The Home Depot and other companies to have representatives who are able to speak on the company's behalf, while also making it clear that they may be offering their own personal opinions.

Caminito Argentinean Steakhouse Case Study: The Steakhouse That Engages Online Everywhere

In the aforementioned three case studies, large companies took the initiative in social media efforts. But social media is still a viable advertising method for small businesses. Indeed, "participation is marketing" can work for the smallest to largest companies. Here's how a brick-and-mortar business with one single presence—a restaurant location in Northampton, Massachusetts—was able to succeed in social media.

One of the partners at Caminito is an Internet Marketing professional by day, with a thorough understanding of the social media space. Justin Levy has used social media effectively in lieu of more traditional forms of advertising, such as newspaper advertisements, travel guide listings, and a prominent feature in the local Yellow Pages.

Levy has used his knowledge to bring Caminito fully into a social media space. The official steakhouse web page (*www.caminitosteakhouse.com*) is updated regularly, but beyond that, it has launched the Prime Cuts Blog (*http://primecutsblog.com*), on which Caminito shares cooking tips on with recipes, gadget reviews, and videos (Figure 4-6). Caminito also has an exclusive YouTube channel (*www.youtube.com/user/primecutstv*), known as "Prime Cuts TV," on which it offers additional "see it yourself" instructional videos. For example, it recently uploaded videos entitled "How to Sharpen Your Knife Using a Wet Stone" and "How to Fabricate a Whole Ribeye." Further, using video-sharing service Viddler (*www.viddler.com/explore/primecutstv*), these videos are published in higher quality on the Prime Cuts Blog.

Caminito has also realized the importance of social media in a long-term "how do people find us?" search engine optimization strategy. Being omnipresent means that people can discover Caminito using a variety of online search phrases. To achieve these results, the restaurant actively encourages patrons on its web page to review the restaurant on business review site Yelp (*www.yelp.com/biz/caminito-argentinean-steakhouse-northampton*; shown in Figure 4-7). It also has tapped into the local youth crowd by creating a MySpace profile page (*www.myspace.com/lasbrasasnoho*), where it posts events and points visitors to the official Prime Cuts Blog.

FIGURE 4-6. The Caminito Argentinean steakhouse Prime Cuts Blog

Justin Levy remains affiliated with Caminito through his personal social media accounts as well (as listed on the Prime Cuts Blog sidebar). He's active on social geo-aware network Brightkite, shares his social media– and food-related bookmarks on delicious, lets people monitor his blog comments through comment-monitoring tool Disqus, highlights his social media accomplishments and personal achievements on his Flickr photostream, aggregates his social media stream on FriendFeed, networks with professionals via LinkedIn, bookmarks websites via social-discovery tool StumbleUpon, regularly interacts with users via Twitter, and broadcasts his upcoming schedule of events at Upcoming.org. Clearly, Justin has shown that there is no shortage of options for the types of interactions he can have with community members.

Of course, Caminito also actively monitors the conversation using Google Alerts. Caminito uses this information to learn more about what people are saying about it and also for competitive analysis to understand what other Argentinean steakhouses are doing in other locations. Using this data, Levy is able to keep abreast of menu changes, new ideas, good and bad reviews, and other tactics Caminito can then use for its long-term growth.

FIGURE 4-7. *Caminito Argentinean Steakhouse's Yelp profile*

To measure success, Caminito monitors web analytics regularly. Based on the data provided, it enhances its social media engagement across several networks or makes necessary adjustments.

In the end, Caminito is seeing much success with its social media marketing initiatives. Its search engine rankings are phenomenally high, which is incredibly important in its market. Its web page often shows up multiple times for a specific search, which ends up bolstering the brand itself and reinforces Caminito Argentinean Steakhouse as the ideal restaurant for the particular query performed. In an interview with Social Media Explorer,[4] Justin says, "We have seen an approximate 30% boost in sales (year to date) in a time where a lot of restaurants are down 10–20%. Not all of that can be attributed to our online presence but I'm sure a good portion of it can."

He's probably quite right.

Reputation Management

You may have spent decades building the empire that now houses your brands and hundreds of thousands of employees. But reputations are precarious; within a matter of moments, your hard work can come tumbling down when a customer (or even a competitor) uses the Internet to tarnish the good name that you have been trying so hard to maintain. Given the proliferation of content on the Internet, one bad story can easily spread like wildfire. Companies that do not react to the firestorm can suffer considerable loss of trust and may even lose brand share.

Social media is a cost-effective and beneficial way to combat these kinds of reputation management fiascos. There are several ways to address reputation management using social media. In Chapter 2, you learned how Electronic Arts transformed what could have been a demerit among its fans into a superb marketing piece that in itself was incredibly viral. As you may recall, a user discovered a software glitch illustrating a humanly impossible feat, and uploaded a video showing this phenomenon to YouTube. EA Sports opted not to ignore this video and responded by making a genuinely realistic illustration of how the software bug was not actually a software glitch after all. With over 2 million views and overwhelmingly positive reactions to the video, EA Sports clearly emerged victorious from what could have been a public relations nightmare.

In the Caminito Argentinean Steakhouse case study, I alluded to a second opportunity for social media to assist in online reputation management. In that instance, you learned that the presence of multiple profiles has made Caminito (and Justin Levy) more findable. Caminito maintains several social media profiles across a multitude of social networks, and a search for Caminito using a typical search engine often pulls up more than just the restaurant home page: the results show Caminito's social media profiles, too. Justin Levy's personal social media profiles are also widespread across the Internet. If you searched for "Justin Levy" today, you'd likely see his social media profiles scattered throughout the first page of the search results. There are probably hundreds of individuals named Justin Levy out there, but it becomes pretty evident relatively quickly that the Justin Levy on the first page of the search results is the man responsible for the success of Caminito Argentinean Steakhouse.

This example illustrates the power of social media marketing for reputation management. A business that may be hurting in the search engine results can easily set up a number of social media profiles across hundreds of social networks. With regular engagement, social media profiles may help influence the search engine results. This strategy works well because most social networks are trusted by nature of their heavy use (and because they are often linked to by other websites and news outlets). Subsequently, internal pages are often trusted because they are associated with the particular social site's domain.

The screenshot of the 2008 Comcast search results (in Chapter 1) shows that user-generated content ranks well in search results. Further, ever since Google introduced "universal search" results, it is evident that social media can even make its way into the more visible links on the search engine results page. A search for "sledgehammer" shows one dedicated YouTube result at the top of the page (and in the center of the page, product listings for the item). A search for "Barack Obama" shows regular results scattered with news results, book results, and most significantly, blog posts.

DEFINITION

What is *universal search*? In 2007, Google decided that the search results were lacking with just 10 plain blue links. Google decided to integrate video, images, blog posts, business data and maps, products, and news results into the search engine results pages. Universal search increases visibility of other possible properties that Google is currently highlighting in its results.

Clearly, social media results are noticeable in nearly every online search. For reputation management, then, social media works very well to combat negative search. Since the majority of social networks allow you to choose your own username, you can claim ownership of the available usernames for your brand or company. Once these social media profiles are ranked within the search results, they can help bolster your online reputation. Use a site like *http://knowem.com* to see which websites you should register your brand name with so that you can then start to claim your reputation on the Internet. You don't have to claim every single social media profile, but you should be visible. Once these social media profiles are registered, be sure to use these social media accounts on a somewhat regular basis (at the minimum) to help create value within your communities. This in turn helps you get higher rankings and prevents others from ruining your good reputation. You will not achieve a high ranking just by virtue of creating the social media profiles; engagement is still necessary. Your social media profile page should show that account activity is being recorded. This is helpful because search engines will crawl (or "spider") more often on pages that are updated more frequently. Thus, the more content you can add to the social media profiles, the more likely it is search engines will notice you. As an added benefit, if you're recognized as a good contributor by members of the communities that you participate in, your social media profile page may naturally be linked by actively engaged social media participants who may also be bloggers or journalists (my Twitter profile page currently outranks my personal blog for numerous search queries).

DEFINITION

What is a *spider*? A spider or web crawler is a scripted program that automatically traverses through the Web and gathers data about websites and web pages. Have you ever wondered how Google, Yahoo!, or your favorite search engine actually knows what text is on a website? Each search engine uses a spider to record all text on a web page. On a traditional website, spidering may not happen as rapidly, because search engines do not know about spiders' existence, but since blog content contacts search engine ping services directly, spiders often come a lot more quickly.

Network Solutions Case Study: Reputation Management by Listening

Several months ago, technology company and domain name registrar Network Solutions was under fire for apparently violating ethical considerations by registering domain names that individuals would seek out for availability. In plain English, users performing simple research on domain name purchases via Network Solutions would not be able to register that domain using another registrar because Network Solutions would immediately claim ownership.

With this questionable practice (which one blogger called "extortion"[5]), Network Solutions had a reputation issue that it needed to address quickly to fall back into the good graces of its customers and prospects. Trust was lost and morale was very low.

After hiring public relations firm Livingston Communications, Network Solutions worked with Chief Community and Social Media Evangelist Shashi Bellamkonda on a viable strategy that would help the company to win back the favor of the Internet-savvy audience. Its strategy required three important steps:

1. Listening, first to monitor online chatter and then to respond on the official Network Solutions blog, Solutions Are Power (*http://blog.networksolutions.com*).

2. Adding value to the community (Network Solutions used its blog to convey thoughts and offer industry-related correspondence).

3. Community participation (communicating in social channels around the Internet on behalf of Network Solutions, leaving the Network Solutions name visible in exchanges).

One of the key contributions to the Network Solutions strategy was the launch of the Solution Stars Video Conference (*http://solutionsstarsvideo.com*), a website featuring a series of videos that discuss how small businesses can benefit from new media tools. Network Solutions interviewed many experts who had extensive involvement in social media on numerous topics, such as how to build a web presence, whether you should start a blog, how to achieve visibility through search, and how social media provides an opportunity to bypass traditional media channels to directly reach stakeholders. This was a perfect opportunity for Network Solutions, because the video conferences provided great value to the community at large (especially a highly vocal community who wanted to share related content), and many experts who were featured talked about it on their blogs.

Over time, the interactions and the video conferences paid off. Over a six-month period, negative comments about Network Solutions decreased and positive sentiment increased (Figure 4-8).

As far as the issue that contributed to the negative emotions about Network Solutions in the first place, Shashi Bellamkonda says that the practice is no longer being employed.

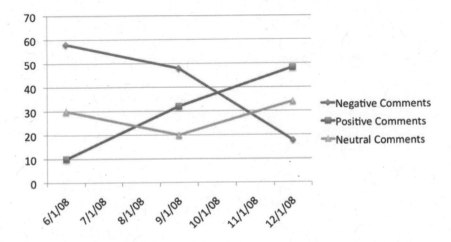

FIGURE 4-8. The change in sentiment of Network Solutions over a six-month period

Today, Bellamkonda contends that social media is a viable two-way strategy that can help all sorts of businesses from the one-man firm to the large corporation. Regardless of company size, social media can facilitate and cultivate strong interpersonal connections. Bellamkonda adds that behind a company's social media identity, one must be genuine and not just regurgitate the same old corporate communication you would see on a traditional media channel.

It's important for Network Solutions to actively monitor the social media space for conversation about the company, but also to respond very quickly, as expectations in the social media space have changed, especially with the rise of tools such as Twitter, where customer service is almost immediate (see Chapter 6 for more information on Twitter). With this type of engagement, Bellamkonda acknowledges that, "people are wowed at the fact that they are getting an immediate response." Bellamkonda says that because Network Solutions is reaching out to customers on blogs, social channels, and Twitter, people are responding more positively toward the Network Solutions brand and are looking at the company in a much better light.

Reputation Management Monitoring: 12 Reputations You Should Monitor Online

It's obvious that you must monitor your reputation online, but what exactly should you be looking for? What other reputations should you watch out for? According to online reputation management expert Andy Beal,[6] your company needs to track the 12 reputations that follow.

Naturally, the course of action for any reputation management missteps will vary based on the circumstances, but by actively observing these reputations, you may be able to prevent issues that could be very damaging. In other instances, the information you discover via monitoring can be very rewarding.

Your name

It doesn't matter if you're a big player in the space or a small one; you should always be aware of what people are saying about you in the media. Plus, you can always link to the positive media mentions on your website to let your visitors observe your accomplishments.

Your company name

Of course, you know this is a given if you're even considering reputation management at all. Listening to what people are saying about you and your company is crucial. Also, consider legacy company names or well-known abbreviations of your company name.

Your brand names

If you're part of a huge company that maintains hundreds of brand names, this may be quite difficult to monitor, but you should follow the more important brands for your business survival.

Your company's executives

Always be aware of what people are saying about your company leadership.

Your company's media spokespeople

Anyone who speaks publicly on behalf of the company should also be monitored.

Your slogan or marketing message

What are people saying about your slogan? Is it being well received? On the other hand, is it being infringed upon?

The competition

What are people saying about your competition? Can you use that information to better your company? Remember, Justin Levy at Caminito Argentinean Steakhouse monitors his local competition to get ideas for his menu and to hear what other people say about other restaurants in the area. Reputation management can still assist with competitive research and analysis.

Your industry

Most specifically, you should observe industry trends and use this information for your own benefit. Perhaps executives are upset that the highly anticipated desk they just bought has a design flaw and a wobbly drawer. Perhaps the mobile PDA many businesspeople received last week has issues with reading external memory modules. Can you learn from this feedback and correct these mistakes to make a better product? Can you use these learning experiences to your advantage? You can also monitor your industry to see new innovations as they're announced. Getting this information early is a great way to stay ahead of the curve.

Your weaknesses

Face it—your product isn't perfect and there's always room for improvement. If you've read the first three chapters, you know that people are talking about you and they'll point out flaws with your brand or products. You can use this feedback to grow.

Your business partners

Do you actively work with a company that is in the news? Perhaps it's for a good thing; perhaps not so much. Would Ponzi scheme scammer Bernard Madoff's investors have been better off if they actively monitored his reputation? Perhaps not, but if you have business partners doing better or worse in the business, you probably want to find out about it. If Google's Q2 earnings do not meet expectations and you are engaging in one of Google's paid advertising programs, you'd probably want to know that it is having a rough fiscal quarter.

Your clients

Hear something great about your clients in the news? Andy Beal postulates that if you directly approach them and wish them congratulations and much success, your retention rate will go up.

Your intellectual property

Any trademarks or copyrights should be actively monitored for infringement abuse or mistaken identity.

Considering a Reputation Management Strategy

You may have been the victim of a reputation management disaster. Now, everyone's eyes are on you. It will be your responsibility to pull yourself out of it, perhaps by communicating directly with the public at large or by addressing the issue silently via social media content promotion.

Bear in mind that you should always take the high ground and handle reputation crises professionally. Remember, you're already being perceived in a negative light and you can worsen public opinion by intensifying the situation with haphazard emotion and poor logic.

When you do approach the community, be honest about the apparent crime you have already committed, then take the next step to explain how you intend to rectify the issue (or inform the populace that you have already taken steps to fix the problem). You should also be available to handle specific complaints personally or you may wish to offer communication channels (that you will have a senior staff member answer quickly) to ease the concerns of your constituents. If the community believes that the discussion will not end simply with a public apology, be sure to give it a place where it can continue the conversation offline.

HOW DO YOU PLAN TO FIX THAT PROBLEM?

Every company has its growing pains. Some companies may be proactive in preventing reputation management fiascos, but sometimes the conflict is unavoidable. Still, stakeholders and community constituents expect a company to own up to its mistakes. When a credit card company has a data breach, cardholders want reassurances that their data is safe. This is exactly what was expected of Twitter when it experienced a data breach in early 2009.

Numerous famous people, such as Barack Obama, Britney Spears, and prominent newscasters, maintain accounts on the Twitter social network. These accounts have thousands of followers. In January of 2009, the accounts of many well-known figureheads were hacked. Instead of reading messages that were relevant to the lives of these individuals, the Twitter followers instead were surprised to read downright inappropriate and unprofessional information about these individuals and guessed that something was horribly wrong. Readers assumed that these Twitter accounts had been compromised, but how?

The blogosphere started buzzing that Twitter was not safe anymore. People were worried that their accounts would be broken into next. But within a few hours, the Twitter team itself acknowledged the problem. In a blog post, the Twitter staff very openly admitted how the accounts were compromised. Twitter used this experience to improve its security, and it did so swiftly.

This level of transparency sets a good precedent of how other companies should react. Days after Twitter responded and addressed the situation, the worries of Twitter users were alleviated. Responding in a timely and proactive manner helped restore faith in the company, and the user base now feels comfortable knowing that Twitter has its back.

Internally, though, you can avert the reputation management mess (for the future) by learning from these experiences. If you have not yet encountered a publicity disaster, it may be a good time to communicate to your staff about the importance of public perception and reiterate that it should tread carefully and offer the best service because consumers nowadays have easy ways of exposing their dissatisfaction with your support or service offerings. Of course, you can always read about the reputation management disasters of your company and use those experiences to learn what not to do in your industry as well.

Summary

The *Cluetrain Manifesto* alluded to the phenomenon that is social media marketing today. Businesses are talking to one another. Markets are conversations and social media empowers consumers to speak directly to its favorite brands.

Participation is marketing. As marketers, we should take an active role in social media channels and engage in dialogue in a very genuine way. Graco Baby, Tyson Foods, The Home Depot, and Caminito Argentinean Steakhouse are four businesses that have been able to illustrate the "participation is marketing" trend with very positive returns for their business.

Social media marketing can also aid in reputation management tactics. There are two ways this can be done. When a company is involved in the conversation, the company itself can help nurture and shape impressions of the public, usually for the better. Network Solutions showed that listening turned negative sentiment into a positive one.

The second way social media can help facilitate reputation management is via social media profile building. Companies can create user accounts named after the brands they need to monitor via social media channels and then use these profiles to boost search engine results. Continuous and meaningful engagement is necessary so that search engine spiders can discover the profile pages, see that there is frequent activity, and ultimately, often by trust, rank these profile pages higher in the search results.

There are 12 reputations that you should monitor consistently so that you can keep abreast of developments and perceptions about your company, your industry, and your competitors. When you're armed with this information, you should be able to easily address most, if not all, reputation management issues.

Endnotes

1. *http://parentpowerindex.com/parentpower-index*

2. *http://technorati.com/blogs/blog.gracobaby.com?reactions*

3. *http://www.toprankblog.com/2008/10/big-brand-social-media-blogwell*

4. *http://www.socialmediaexplorer.com/2008/10/27/social-media-for-small-business-caminito-argentinean-steakhouse*

5. g (*http://www.billhartzer.com/pages/network-solutions-registering-domains-after-availability-lookup*)

6. *http://www.marketingpilgrim.com/2008/04/online-reputation-monitoring-campaign.html*

CHAPTER FIVE

Using Blogs to Communicate, Influence, and Learn from Your Constituents

In the last 10 years, blogs have grown to be more than just personal diaries; today, they are vehicles of change within organizations and show that many new companies are embracing the two-way communications culture. Blogs are excellent communication tools—you can start your own to convey your company's feelings to the wider public, or you can reach other bloggers who might blog about your product. Today, some very powerful blogs have been able to successfully connect consumers directly with service providers.

A Short History of Blogging

A *blog*, which is short for *weblog*, is a website that is usually maintained by individuals or groups, and more recently, businesses, that offers commentary and ideas for a larger audience. A typical blog features an entry, often sprinkled with graphics and video, followed by comments by readers. The entire layout of the blog is displayed in reverse chronological order with the newest entries posted first.

Blogs are different from static websites because they incorporate social elements. Most, if not all, blogging software includes RSS (real simple syndication), which is a format that allows content to be accessed within many publications. Blogs often invite users to comment, and an ensuing dialogue can occur, which can generate hundreds to thousands of responses.

DEFINITION

RSS is a popular web format for publishing frequently updated content, such as blog posts and comments, news articles, and podcasts. RSS feeds are documents that contain summaries of relevant content from websites. Programs that understand the RSS protocol allow people to keep abreast of the latest happenings on their favorite websites.

The History of Blogs, 1998–2009: Who Is Writing and Who Is Reading?

Blogs are now over 10 years old.[1] Technorati, one of the premier blog search engines, tracks over 133 million blogs. Former chairman of Technorati, Dave Sifry, released a "State of the Blogosphere" report in September of 2008[2] that culled numerous website statistical data, indicating that blogs have between 77.7 and 94.1 million unique visitors in the United States alone, which amounts to approximately 50% of all Internet users in the country.

In an earlier report,[3] Sifry provided more eye-opening information about the nature of blogs: 120,000 blogs arise daily. Each second, 1.4 blogs are created from scratch. This amounts to 1.5 million posts per day, or 17 posts per second. *BusinessWeek* contends[4] that there are 13 million regularly updated blogs and argues that the blog is "simply the most explosive outbreak in the information world since the Internet itself."

With these statistics, it should be apparent that there's a huge market of individuals that you can tap into by blogging. But who are the readers, and is it even worth the hassle? In the 2008 Technorati blogosphere report, it is evident that brands are very heavily talked about:[5]

> Whether or not a brand has launched a social media strategy, more likely than not, it's already present in the Blogosphere. Four in five bloggers post brand or product reviews, with 37% posting them frequently. 90% of bloggers say they post about the brands, music, movies and books that they love (or hate).

Even earlier statistics support the notion of the power of blogging. As early as 2004, Pew Internet reported[6] that 32 million Americans, or 25%, consume newsworthy content via blogs. Another Pew Internet report[7] released that year found that nearly half of Internet users have contributed their thoughts and feelings to the online world. The findings suggest that content creation via blogging and social media community engagement is preferred by more than 53 million Americans.

Two years later, in 2006, a third Pew Internet report[8] discovered that bloggers are the new influencers and authors of the online world. At that point, 54% of writers were found to be using only the blog, with no other form of media, to share their thoughts with the rest of the world. The study showed that 57 million Americans were actively reading blogs via direct methods or through RSS.

If you're unfamiliar with the blog, you may assume that it's just an online journal. Today, it's estimated that only 5% of blogs that exist are dedicated solely to a business purpose. At the same time, this means that it is a great opportunity for others in business to use the blog to market their products and businesses and to establish themselves as authorities in their niches.

But is that all? Other statistics[9] show that 58% of individuals have been turning to the Internet to address problems and issues. As Figure 5-1 shows,[10] a large percentage (53%) of individuals prefer to consult with professional advisors, many of whom may already have established an Internet presence.

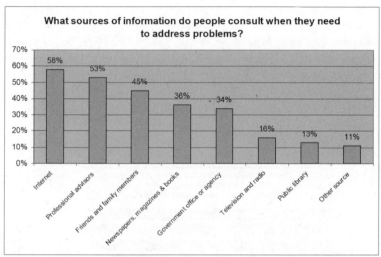

Source: Pew Internet & American Life Project and University of Illinois Libraries Survey. June 27-Sept. 4, 2007. N=2,796 for entire sample, including an over-sample of 733 "low-access" respondents. Margin of error is ±3% for entire sample. For internet users N=1,702 and margin of error is ±3%.

FIGURE 5-1. Who and what do people consult with to address problems?

These statistics are eye-opening. Your blog and your voice can begin to lure customers into your brand and get them talking about your business. Blogs give you power and exposure that you may never have dreamed possible before. With blogs, you can learn what people are saying about your company; actively encourage a dialogue that can improve your business and its products and services; and create a sense of satisfaction by empowering your customers, allowing them to offer feedback, and encouraging them to respond to concerns on your website. In essence, blogs provide new trustworthy relationships with your customers.

Blogs As Online Influencers

Today, bloggers are perceived as a lot more than just writers. In a world where so many people turn to the blog for advice, feedback, input, and commentary, bloggers are finding that not only are they writing for themselves, but many are also assuming the role of the journalist. In the past several years, a new style of news reporting has emerged: blogs are breaking news faster than traditional media outlets, and this is often because bloggers do not necessarily need to jump through the same editorial hoops as regular print publications. Today, due to the swiftness of news breaking via blogs, traditional media outlets are using blogs to supplement and support their own research.[11] Indeed, the Internet has already taken over newspapers as the optimal news source,[12] and traditional media journalists are finding that in order to thrive in this increasingly digital world, they must turn to the blogosphere for the latest news.

User-generated content is also here to stay. Figure 5-1 shows that 58% of polled participants consult the Internet for advice. Whether that means they refer to blogs to read about products or to engage in the discussion, Internet users everywhere want to have a voice in the process. Many people today are active participants on already established blogs. An early 2008 study indicates that bloggers are considered credible sources of information about companies, but Figure 5-2 also shows that blog readers consider CEOs, regular employees, nonprofit representatives, healthcare specialists, academic resources, financial industry analysts, and people like themselves as powerful, credible sources of company information. Would you find it shocking if those individuals also maintained blogs?

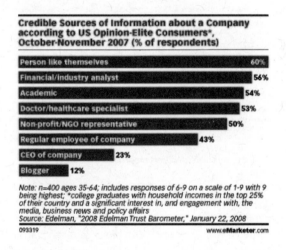

FIGURE 5-2. Credible sources of information about a company

As shown in Figure 5-2, eMarketer has found[13] that 60% of shoppers are influenced by their peers. This large percentage speaks volumes about how powerfully a discussion on your website can influence your brand and products. Like-minded individuals will often flock to common ground (such as a blog or user-generated website that addresses their wants and needs) to engage and to discuss the products, often with the goal of offering feedback for improvement. Beginning a dialogue on your website increases user trust and belief in your product. It indicates that you, as a producer or service provider, are interested in knowing what others have to say about your product and are open to feedback. As a tool for direct communication, blogs help reinforce faith in your brand. The perception of listening and responding is an incredibly powerful trust-instillation mechanism that you shouldn't ignore.

If you're not yet a blogger in your industry, someone else will seize the opportunity. If you don't blog about your product, someone else will talk about it. According to Technorati's 2008 report on the impact of brands on blogging,[14] day-to-day experiences with customer service are fodder for blog content. By starting your own blog, you bring the conversation to territory where you open the floor for your consumers and address other concerns posted throughout the Internet.

Another benefit of blogging is that the more prolific bloggers who provide valuable and consistent content are often considered experts in their subject matter. Not only does this help the blog itself become more of a known brand, but it also gives the blogger an opportunity to "coach" traditional media journalists. Today's news articles feature more quotes and feedback from bloggers than ever before. This is a growing phenomenon; if a journalist wants expert advice, she does not need to go farther than her search engine to get the name of a credible source on a particular topic.

THE POWER OF THE BLOG: TALES OF A STOCK MARKET PLUNGE

As more and more journalists use blogs as sources for verified news, consumers are tuning in to those same sources as credible news providers. But when news breaks that is not accurate, the results can be disastrous. In 2007, popular gadget blog Engadget reported[15] that Apple planned on delaying its Leopard OS as well as the highly anticipated iPhone. The unsubstantiated rumors cost Apple an astounding $4 billion dollars[16] when its stock plummeted after the news broke. If anything, this story goes to show that the word of the blog is powerful and can make or break a business. It also illustrates another important point: Engadget was the only news source to post on this Apple finding, and it single-handedly caused a significant drop in Apple stock. This demonstrates that blogs are becoming equated with trustworthy news sources and that people are relying on accurate reporting of information that can influence critical business decisions.

A blog can be a tremendous asset in building a brand empire. With the right content and right approach toward your audience, you can develop strong links that will help you build a solid reputation online. As you continually churn out more content, you will grow a base of readers who will check your site frequently (or subscribe) and build search traffic, and as a result, you'll establish thought leadership through the gain of significant links from other writers and bloggers.

How Blogs Are Consumed

There are a variety of tools available that can provide you with the latest blogging updates.

Direct Hits

A blog is nothing more than a website, and thus many people will get their news from blogs simply by navigating directly to its home page and reading the content. While blogs can be consumed via other methods, the ongoing relevant conversations will often be held on the actual website.

RSS

Another valuable way to consume blogs is via RSS. A number of blog software applications incorporate RSS functionality. Additionally, porting Google's Feedburner (*www.feedburner.com*) options on top of your RSS feed may be ideal, as it is a powerful suite of tools related to RSS that affords bloggers the opportunity to integrate advertising in their feeds, see detailed statistical information about readers and subscribers, and optimize the feed itself.

With RSS functionality intact, readers can enjoy regularly updated content via sites such as Google Reader (*http://reader.google.com*) or Bloglines (*www.bloglines.com*). These online RSS readers often have a panel that displays the online publications that the user has subscribed to, and another panel that displays the content. Figure 5-3 shows just how much opportunity there is to read and review a lot of content very easily.

Alternatively, users can view RSS content on personalized web pages, such as My Yahoo! (*http://www.my.yahoo.com*) or iGoogle (*http://www.igoogle.com*), or via services such as Pageflakes (*http://www.pageflakes.com*) and Netvibes (*http://www.netvibes.com*). These options are preferable for those who like to see content that is custom-tailored to their needs, with their email information in one pane and weather information in another. News stories are often displayed on these pages as well. These new stories, believe it or not, are being pulled via RSS.

Blogs via Email

Users can also consume blogs through email alerts, which can be set up via Google Alerts, as discussed in Chapter 2. They also can be delivered in the form of a newsletter, which can be activated with Feedburner or a dedicated service like Feedblitz (*http://www.feedblitz.com*). In other words, subscribers do not necessarily have to visit any particular web page to read a blog; content from blogs may be delivered directly to the recipient's inbox.

FIGURE 5-3. Google Reader offers detailed information, accessed via RSS, with a user-friendly pane and short introductions to content

A Beginner's Guide to Blog Platforms

There are a number of blogging platforms available, and some provide features that are not offered by others. Keep in mind that you should choose your blogging platform carefully. You may choose initially to go with a free solution, but in the future, your blog and brand will grow and you might be required to migrate to a different type of platform and host. While there are tools that make this transition relatively easy, you should avoid doing so, if possible. If you're really enthusiastic about blogging, the ideal solution is to flesh out the details now, because the costs may be aplenty later (such as the time investment involved in extensive website development and migrating from the previous website provider).

Features and Functionality

Blogs are probably the most common publishing platform for content producers, because it is just so easy for website publishers to create content quickly and professionally. With most blogging platforms, it is extremely simple to add links and images, and the content creator does not need to possess considerable knowledge of HTML. Numerous blogging platforms boast comprehensive WYSIWYG (what you see is what you get) editors.

DEFINITION

WYSIWYG is a feature that shows the end result of a processing task as it is being produced. In blogging-speak, a WYSIWYG editor lets you see specific formatting (bold, italics, links, images, etc.) without the actual HTML code. For those who are not familiar with HTML, WYSIWYG editors are extremely helpful.

Most blogging platforms also integrate RSS without requiring you to set up syndication channels yourself. One of the biggest benefits of blogging software is the capability to contact a variety of servers (specified by you and/or provided by the blog software) so that search engines can find content updates quickly. This action, called *pinging*, occurs when publishers push out a new update. The benefit of pinging mechanisms is that content is spidered extremely quickly; I've witnessed content being spidered from even the smallest of blogs within an hour of publication, thereby helping you get high search result rankings much earlier than other "static" content. Pinging also updates RSS aggregators and shows your readers the most recent content. Thus, blogging is a powerful and instantaneous publishing tool that does not require you to refresh that web page frequently for updates.

In stark contrast, static websites do not come with RSS features or pinging services. If you update content, you may still have to wait until a search engine spider comes along to your website and then crawls your content. This process is not very fast. With blogging software, the content crawl time is expedited tremendously.

Blogging Platforms

There are numerous blog platforms available. Some are free but require you to maintain them. Others are also free but are hosted on a centralized server and give you very little administrative control over the backend. Others are full-fledged paid solutions that boast security updates, and are not as vulnerable to exploits because they are not very dominant players in the blogging market. Here are some commonly used blogging platforms:

WordPress (http://www.wordpress.org)

WordPress (Figure 5-4) is by far one of the most popular blogging platforms available. It's fully customizable and has hundreds of thousands of developers behind its plug-in[17]

and theme[18]development. The unhosted WordPress version is a download that you install and maintain on your web host, and once you install it, it is completely modifiable. In fact, numerous websites are hosted using WordPress software and yet they do not even look like blogs.

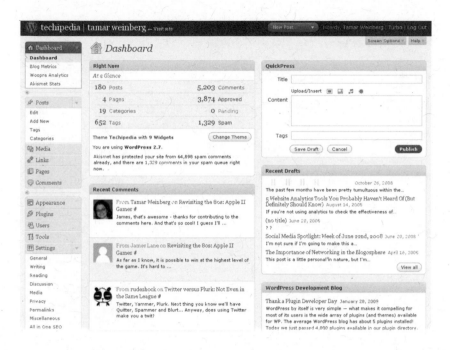

FIGURE 5-4. A very detailed WordPress dashboard combines all blog features into one place

WordPress (http://www.wordpress.com)

Another version of WordPress, not to be confused with the previously discussed version, is a centralized and hosted solution on WordPress's official site for those who may not have their own domain and hosting solutions. Interested parties can sign up and have a blog running within seconds. Because this solution is hosted on WordPress's servers, you have a lot less flexibility in customization, plug-ins, and theme support.

MovableType (http://www.movabletype.com)

MovableType (Figure 5-5) touts itself as a professional platform and is used by some of the most powerful bloggers. It requires a local download and must be uploaded to your web host. With MovableType, users can maintain multiple weblogs, which is especially useful for those who manage several blogs, from a single administrative interface. It also offers customizable templates, in-depth user management, and more.

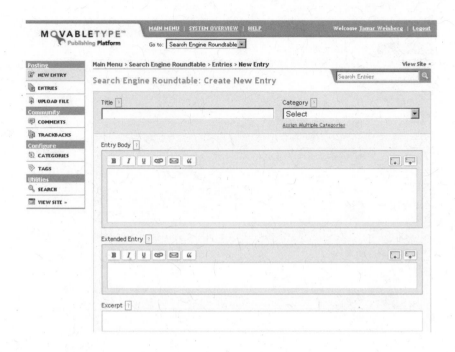

FIGURE 5-5. The publishing backend of MovableType

TypePad (http://www.typepad.com)

TypePad is maintained by the same company that runs MovableType (Six Apart Ltd.), and is the hosted version of the software (it runs on a centralized server). The bells and whistles are very similar to the MovableType feature set, but TypePad is not free. The cost for TypePad[19] ranges from $4.95 per month for the basic version to $89.95 per month for the business-class version. In the basic version, a single blogger can maintain one blog with a maximum of 100 MB storage space. It does not allow you to map your URL (which would probably be similar to *yoursite*.typepad.com) to your own domain name; that branding and personalization feature is offered at the next pricing level.

Blogger.com (http://www.blogger.com)

Google maintains its own blogging software with Blogger.com. This is a hosted solution that is available on a subdomain on blogspot.com. Blogger.com is a very simple solution, but is not very modifiable.

ExpressionEngine (http://www.expressionengine.com)

ExpressionEngine is a comprehensive publishing platform with full customization and control, first-class support, a powerful template engine, an integrated mailing list, plugins, built-in utilities, and more. ExpressionEngine is a hosted solution, and you must have a compatible web server for the software to work. A basic version of the software is free, but the more powerful and recommended versions are $99.95 for the personal version and $249.95 for the professional version.

What Software Should You Use?

Do you know where you want to host your blog? Before you make this decision, whip out a pen and paper and write down the answers to a few questions:

Determine your goals:

- Is this something I see myself doing for the long haul?
- Will this blog be used for myself or for my company?

Determine your budget:

- Do I have the budget for hosting and domain name registration?
- Do I have the development budget for a customized design?

Assess your technological skill set:

- If a high-priority security update is released, can I fix it immediately or do I need to get the IT department to patch the vulnerability?
- If the blog stops working properly, can I fix it?

Based on the answers to these questions, you can get an idea of the kind of software you should choose.

Depending on your needs and requirements, you may want to go with blog software that is simple to manage and does not require much IT overhead. If that's the case, a hosted solution may be preferable. On the other hand, you may want to have full control over the software and should opt for one of the other tools. You may also want to monetize your blog via advertisements on the sidebar. A hosted solution may not accept advertisements, so your desired blogging software may be one that you will manage, maintain, and customize yourself.

There are some other considerations regarding blogging platforms:

- If you or those composing blog posts on your blog are not HTML-savvy, you may want to consider a solution that boasts a powerful WYSYWIG editor.
- The most powerful fully managed solution is the unhosted version of WordPress, because it has thousands of plug-ins and themes. This would likely be the preferred blogging platform for monetization.
- The more popular and free downloadable platforms (WordPress and MovableType) release updates on a pretty regular basis (monthly or bimonthly) because they are so widespread that they are vulnerable to attack and can be easily exploited. Likewise, the plug-ins on WordPress are also vulnerable to attack, but in more recent versions, plug-ins can be

updated automatically from the administrative backend (the dashboard) without the need to use FTP.

- If you expect a surge of traffic, perhaps because you intend to use the blog as a platform for viral marketing campaigns that may bring hundreds of thousands of visitors in a short period of time, you need a very powerful web host to accommodate the traffic. WordPress creates dynamic web pages via PHP and MySQL code, but if you have a lot of unexpected visitors very quickly, you may experience a system crash on your own server. If you host with WordPress, this is not a consideration, since system resources are maintained by Automattic, the company that owns and operates WordPress.

Writing for Blog Audiences

If you're ready to begin writing your blog, you need to consider crafting a voice that is believable, genuine, and transparent. The most successful blogs offer a level of transparency between the reader and the writer. Instead of merely providing a corporate manuscript, blogs help readers view individuals in corporate positions as humans. Blogs have a different overall feel, and convey your thoughts about your business in a more informal and humanizing way, thus ensuring that posts are well received. The right corporate blogging types have voices that are heads above the crowd.

In reality, every big company has a website. Among these, however, how many big companies engage in blogging or social media to communicate with their potential and current shareholders? Not very many.[20] Blogging, then, is a great way to influence hundreds of thousands of web surfers who may be interested in your product offerings. Above all, blogs are very powerful because they allow you to be more direct with readers and build real relationships with them.

Most bloggers are looking to open the doors to conversation, and participants on comment-enabled blogs feel more welcome and appreciated, especially when the company is perceived to be listening (and senior executive officials are engaging in the ongoing conversations). Those who participate in the conversation may feel as if they're contributing to the overall business strategy. Blog comments empower users and make them feel good about their investments in the product or service offering. Giving readers a sense of entitlement by allowing them to participate has often given community members a strengthened belief in the business model. Not everything has to feel "corporate," after all.

Crafting the Blog's Voice

When considering the voice of the blog, don't think too long and hard about a detailed strategy. In many ways, the blog voice doesn't need to come from much farther than the heart. Beyond thinking about your sales goals, think about how you can relate to your community; your "audience" may not necessarily be buying your products (yet), but with the proper tone, you can convert these people into readers, subscribers, or buyers. When speaking to your readers through the blog, ignore the corporate jargon and put them first. Speak to them on a human level. Use the blog as a vehicle to convey ideas to your readers, but remember that it's very important to appeal to them emotionally as well.

NUTS ABOUT SOUTHWEST CASE STUDY:
A CORPORATE BLOG DONE RIGHT

One of the most highly acclaimed blogs for achieving this balance between appealing to readers emotionally and conveying company goals is Southwest Airlines's corporate blog, Nuts About Southwest (*http://www.blogsouthwest.com/blogsw*). In honor of Asian/Pacific American Heritage Month, one employee talks about her youth as a Vietnamese American. In another post, a father reflects on his policeman son's duties to safety while recounting the recent murder of a fellow policeman. In another, the Southwest Airlines team shares its experience after attending the Grand Ole Opry Verry Merry Christmas Parade. And in a post from Valentine's Day, the airline integrates a story of love with how it operates the business. It is evident that Southwest Airlines is not just thinking about its own goals, but also about its readers. As a result, the readers feel more part of the "Southwest Airlines family." The blog has been an incredible success for the company and it even won an award as the best blog from 2007.[21]

If nothing else, the Southwest Airlines blog should exemplify how blogs can be and are successful within the corporate world. This particular blog is a success because it is personable and the writers are directly interfacing with customers (and valuing them by allowing for conversation) while incorporating human elements, particularly emotions. In the case of the father reflecting on his son's work as a police officer, the story itself is emotionally touching and gripping. The blog acts as a powerful conduit to bring in additional readers who may want to read about personal heartwarming stories and to learn about the business—because, after all, that is still one of the key goals of this blog. The Nuts About Southwest blog may be a corporate blog, but it reflects the attitudes of employees so naturally that it does not feel corporate.

Techniques and Tactics

When you compose a blog post, appearance is everything. Blogging is still very much part of the social media umbrella, so to ensure that blog posts are well received among social media influencers and other readers, you should employ the following techniques and tactics.

Use visual elements to capture attention

Text alone doesn't work anymore; that's a strategy that worked in the traditional media days, but today, it is far more important to aesthetically dress up blog posts with video, images, icons, graphs, charts, and other visual enhancements.

Keep it clean and to the point

The writing style is important. Break up posts into concise and readable paragraphs. Don't ramble and don't deviate from the main idea. Blog posts are not formal essays, so there is no need to overwhelm readers with irrelevant information. As such, it may be helpful to break blog posts into an introduction, a body, and a clear conclusion. Today, due to information overload, your readers are probably skimming your articles[22] and not reading them in their entirety. Therefore, consider using bold or italics to stress key points. When you finish writing your post, don't publish it immediately. Take a breather and then proofread it before it goes live. You'll be happy that you did.

Link to appropriate sources

As blog Problogger.net suggests, "Don't be an insular blogger."[23] Link generously and appropriately to outside sources. This, in turn, generates trackbacks in the blog software that can build up a powerful network and can get you noticed as a blogger.

> ### DEFINITION
>
> What is a *trackback*? A trackback or a pingback is a notification that alerts web publishers when someone links to their articles. Typically, this does nothing but generate a link back to the original post for the reader to learn more. However, it is a powerful way to learn exactly who is linking to you and to let blog readers know that the article in question is being discussed elsewhere on the Web. Many blog applications support trackback functionality by default with no extra intervention from the designer or blogger.

Write powerful headlines

Powerful headlines can make or break attention spans and can act as hooks to draw your readers in or to push them away. Brian Clark of Copyblogger.com remixed many headlines in a series that he published on his blog. In one instance, he transformed former title "The Web Next Revolution" into "Why the Next Web Will Be Smarter Than You," arguing that the new headline presents a challenge to readers and raises the curiosity factor. Your headlines should provoke, confront, and speak directly to your readers.

Provide reader-friendly lists

Readers today are skimming online content more than ever, so creating blog posts of lists, if done in moderation, can keep a reader entertained. The reason lists are useful is because they are more easily digestible than the standard paragraph format. Another reason is that most lists are valuable and are often bookmarked, thereby accomplishing an important goal of social media marketing.

Write informative how-to articles

Blogs are usually informative in and of themselves. However, most people will find you because they are searching for information on how to achieve something. Copyblogger author Brian Clark argues that, "If you think you're giving away too much information, you're on the right track."[24] Never be afraid to provide as much information as possible, because this compels your readers to read and learn more (and even then, you're still the expert since you often speak from direct experience, and people will still use the services you are offering to sell). In turn, you establish yourself as an expert, and without a doubt, you will gain more customers due to your in-depth knowledge.

Use storytelling to your advantage

Lure in readers by telling a story about yourself. Recall that this is how the Southwest Airlines blog grew. Appeal to your readers on an emotional level and use your blog to present yourself as human and approachable. The more open you are about yourself, the more likely it is that your readers will be receptive to you and open up to you as well. In a kind of "pay it forward" way, blogs are creating a means for people to be a lot more open about who they really are. There are obviously limits to this, and each blogger will need to create these boundaries and adhere to them. Some bloggers may talk about politics, but never blog about their family lives. Others will talk about their families, but never talk about their professional relationships. Each blogger has her own style and will make it work based on her comfort level and user feedback.

Use interviews to encourage engagement credibility

Interview posts can work very successfully in many ways. You can run a series of interviews with a variety of experts on a topic, or you can engage your readers by asking questions and letting them contribute, regardless of their level of expertise. Most interview posts will generate a lot of traffic and links, and many individuals are quick to share information about themselves in this digital era. At the end of the day, blogging is about you, me, and everyone else involved, and the communal effort is typically a great way to create awareness.

Write reviews of relevant products or services

Talk about products that pertain to your readership, especially if they make your audience's lives easier. If you have had a particularly empowering experience with something that you think can benefit your readers, share it with them. Additionally, if the service you are advertising requires payment, you may be able to monetize off the links by using affiliate programs. A good example of this can be seen on Jeremy Schoemaker's blog, shoemoney.com, where he occasionally uses his very popular blog to promote relevant books for his readers. In his posts, he links directly to an Amazon.com affiliate page and gets commission from every sale sourced to his website.

TIP

To make money from affiliate links, you will need to join affiliate programs. Amazon.com's Affiliate Program is located at *https://www.affiliate-program.amazon.com/*. There are dozens of other affiliate programs that you can explore and investigate, and sometimes, your favorite product or service will also have an affiliate program that entitles you to high commissions from successful sales.

Use regular features to build a following

A regular feature that covers a specific topical area can increase traffic. You may have a section called "Ask Mr. X," where one of the executive personnel takes readers' questions and provides real, genuine answers. Perhaps you want to entertain the possibility of airing a videocast on your website twice a month that summarizes the biggest developments in your industry. You may want to publish a book review every Wednesday. Regular features drive user expectations for specific types of content. Foster a sense of expectation in your readers by providing certain content on a certain day; more successful features will likely bring a boost in readership and more readers over time.

Listen to your readers

Once you start constructing readers' expectations through your first few posts, you can use their comments to gain insight into what they think of your blog. Use their feedback as advice on how you should proceed. There really is no wrong way to blog, but avoid making the blog sound like you are providing carefully crafted marketing messages that appear to have been run through numerous internal departments for approval. The blog should always be real. The Southwest Airlines blog shows exactly how far and powerful a real blog can go to achieve victory.

Don't abandon your readers

If you cannot update posts for a prolonged period due to holidays or personal leave, be sure to inform your readers of your unavailability. Your readers will not stick around if you are perceived as being inexplicably unavailable to them. Instead, invite readers to guest blog for you for a post or two, or ask experts to share their thoughts. Let readers have a voice and empower your community. Fortunately, most bloggers who partake in the spotlight will not expect compensation; they just want to be heard as part of the community.

Content Strategies for Bloggers: How to Find Inspirational Content

You're ready to dive in and write for your blog, but you feel that your ocean of ideas will soon dry up into an oasis. Without a constant influx of blog posts, you can lose your audience. How do you get inspired with ideas for fresh content?

There are hundreds of places online where you can get ideas for current and future blog posts. Blog directories and topical websites, news articles, and even regular social media are great ways to find good content for your site.

Google Alerts

Google Alerts is beneficial for conversation monitoring and reputation management, but it is also wonderful for inspiring you with new content, as it is posted and thus discovered by Google's spiders on the Internet. If you subscribe to alerts for a specific topic, you will likely get a significant amount of email messages and thus will have no shortage of ideas to write about.

Other blogs

With millions of blogs already in existence, you can bet that there are already other blogs elsewhere on the Internet on the same subject matter as yours. Use those for inspiration. One note of caution: many blogs exist simply to regurgitate blogs for search traffic; to stand out from the crowd, add opinion and insight into your blog posts. Write detailed commentary and cite your sources. Wherever possible, link to associated blog posts.

> ### TIP
> For additional inspiration, check out Alltop (*http://www.alltop.com*). Known as the "online magazine rack" of popular topics, Alltop features thousands of blogs on hundreds of topics and is a great place to find related blogs for insights (Figure 5-6).

Blog search

Technorati (*http://www.technorati.com*) covers popular trends within the blogosphere and is a useful search engine to find out what's popular right now. Alternatively, Google Blog Search (*http://www.blogsearch.google.com*) (shown in Figure 5-7) lets you search for blogs and can be narrowed down to most recent posts, which is helpful to pinpoint blogs that were published with breaking news and information.

Social bookmarking

One of the hidden gems within social discovery bookmarking site StumbleUpon is the ability to search based on specific tags. You can discover sites based on a search phrase (on the StumbleUpon Toolbar, click All→Search) or on tags (by accessing *http://www.stumbleupon.com/tag/searchphrase*). Similarly, delicious.com lets you search for tags using *http://delicious.com/tag/searchphrase* or *http://delicious.com/popular/searchphrase*. (Social bookmarking is discussed in detail in Chapter 9.)

FIGURE 5-6. Alltop lists thousands of blogs in a specific subject area

News stories

Another way to keep the content fresh and frequent on your blog is by including breaking and spreading news. Search for relevant news in the topic you specialize in on any news site. Additionally, some sites aggregate the most newsworthy content. If you blog about technology, for example, you may find Techmeme (*www.techmeme.com*) valuable (Figure 5-8); political aficionados may prefer Memeorandum (*http://www.memeorandum.com*). Ballbug (*http://www.ballbug.com*) is for those interested in baseball, and WeSmirch (*http://www.wesmirch.com*) features celebrity gossip. These sites aggregate news sources and social reactions through blogs, and algorithmically determine which information to post (with some human intervention).

Google Trends and Insights

Google Trends (*http://www.google.com/trends*) and Google Insights (*http://www.google.com/insights/search*) show you what people are searching for and emphasize exactly what is popular throughout the world. If you perform a search (with commas), you can compare two or three search phrases at once. Detailed data (including the most popular time for a particular search), the countries that search for the term most frequently, and related search phrases are then provided, as shown in Figure 5-9.

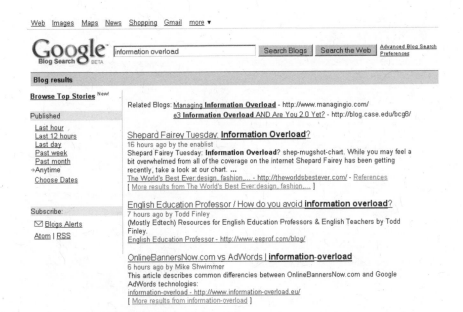

FIGURE 5-7. Use Google Blog Search to find recent articles on a specific subject

Blog Enhancements That Work

If your blog will encourage commenting and community engagement, it's time to emphasize your blog as a social platform. You can add interactive widgets to your blog's sidebar, for example, to let people network with you and see your readers. You can evoke emotion through compelling visuals that are either stationary or moving (photography or video). You can also turn your blog into a social platform by integrating strong community-oriented commenting tools. Whatever the case may be, you're writing for your audience and not just for yourself, so keep your readers happy by involving them in the discussion and being open to their questions and feedback.

Widgets

One way to enhance your blog is to incorporate widgets into it. Two of the most popular widgets among bloggers for community engagement are MyBlogLog and CloudShout. Both of these are interactive widgets that inform the blogger of recent visitors and even provide detailed analytics packages. MyBlogLog has community pages that allow users to communicate offsite and associate themselves with their favorite blogs. CloudShout (Figure 5-10), which was launched much later, takes the MyBlogLog functionality to the next level by allowing users to join blogging communities, install third-party applications to extend the blog's functionality, and view mashups from Twitter, Flickr, YouTube, and other popular social media sites.

FIGURE 5-8. Techmeme aggregates frequently discussed news from the technology industry

Visuals

Do not forget the importance of involving visual elements in your blog copy. Lure your readers in with appropriate eye candy. An amusing image or detailed chart may cause your readers to stop and take a second glance. The right pictures can truly make a solid first impression on a new reader and keep him entertained. Two great places to find images that are acceptable to use (without copyright concerns) are Creative Commons licensed images on Flickr and free stock photos through a site like EveryStockPhoto (*http://www.everystockphoto.com*).

> **NOTE**
>
> Images that are licensed with Creative Commons give publishers the opportunity to share content, such as images, with less restrictive copyright enforcement. If you locate a Creative Commons–licensed image that you want to use, the original artist will likely request an appropriate credit. Be sure to link to the original photograph somewhere in your blog post. For more information about Creative Commons licensing, see *http://creativecommons.org*.

When you do use images, use an image editor to resize the image. Don't just let your blog software force height and width constraints on an existing photograph, as the image will not

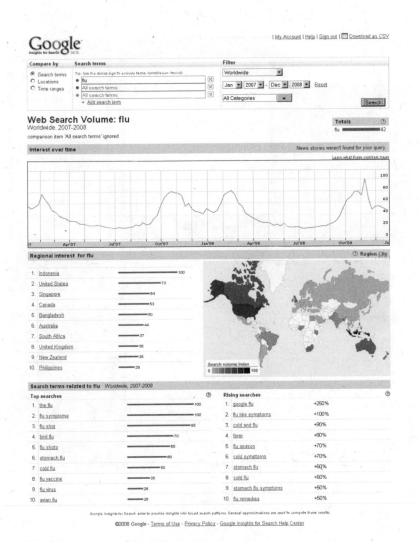

FIGURE 5-9. A Google Insights search for flu data

look as nice and will take longer to load. If absolutely necessary, use a small image on the blog post and then link your readers to a full image elsewhere.

If your hosting plan does not allow you to upload images, use a photo service like Flickr, ImageShack (*http://www.imageshack.us*), or Photobucket (*http://www.photobucket.com*) to host these images. Flickr is a favorite among many bloggers and photographers for its social media benefits and active community; it will be discussed in Chapter 11.

FIGURE 5-10. CloudShout can show you recent visitors and is an interactive communications application with draggable windows

Videos

Since video also elicits an emotional response from viewers, its helpful to incorporate it into blog posts where appropriate. If you create a video, be sure to provide a transcribed version for search engine optimization purposes. Many sites do not take advantage of this, even though all could benefit from the search traffic. There are a myriad of video-sharing sites out there, from YouTube to Blip.tv to Vimeo to Jumpcut to iFilm (now Spike) to Viddler. It never hurts to upload your video everywhere for maximum exposure. Once you upload to one of these websites, you will be provided with an embed code that you can place within your blog.

Video comments make it easy to add human elements into your website . A site called Seesmic (*http://www.seesmic.com*) is empowering bloggers to converse with video. Seesmic has released a plug-in for WordPress to make it easy to embed video comments.

Comments

If you want to engage readers across multiple websites, instead of allowing them to write comments on your blog using its built-in commenting software, you may want to consider integrating Disqus (*http://www.disqus.com*). Disqus (Figure 5-11) is growing in popularity with tens of thousands of blogs currently engaging in the technology. Disqus typically replaces the blog comments of your standard blog installation, but allows for threaded comments, conversation tracking, and more.

Involve Your Audience

When you're a blogger, your audience is your biggest asset, so it's important to encourage the audience to participate and also to enlist your readers' assistance. There is simply no shortage of creative opportunity in terms of keeping the readers entertained. These are just different starting points, but as you find your way and understand the users' responses, you will see which strategies perform better for your audience.

Ask your readers

Are you running out of ideas? You can always run a column in which you ask the readers a question and invite them to share their answers with you (and consequently, the rest of your audience) through the comments. Depending on the answers provided, this may be a perfect opportunity for you to expound upon some of the ideas your readers have shared and turn that into a separate blog post (or series of posts, if there are many brilliant comments). On the same note, you may want to turn this strategy around and prompt the readers to ask *you* a question. It may be the perfect opportunity to share the benefits of a product launch, to clarify some company policy, or to go on the humanizing level and talk about yourself.

Utilize contact forms

When you launch your blog, make sure it comes equipped with a working contact form so that your readers are able to reach you in some way. Many plug-ins are available that will allow you to set up a contact form swiftly and painlessly. Contact forms are great for users to ask questions or for service providers to send you materials and story tips that may be relevant to your readers. Note that if your blog gains momentum, you may get a lot of irrelevant pitches (which is indicative of how many public relations firms are totally not social media–savvy). Further, contact forms should always have some sort of spam protection, such as a CAPTCHA (Figure 5-12) or math question, so that you are not bombarded by bots.

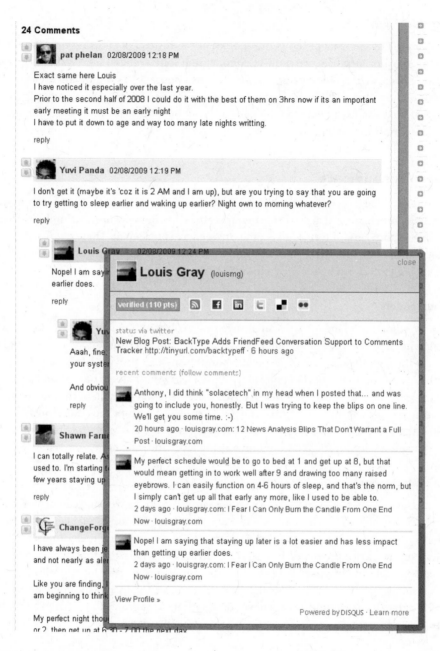

FIGURE 5-11. *Involve people in a deep discussion using the interactive Disqus commenting tool*

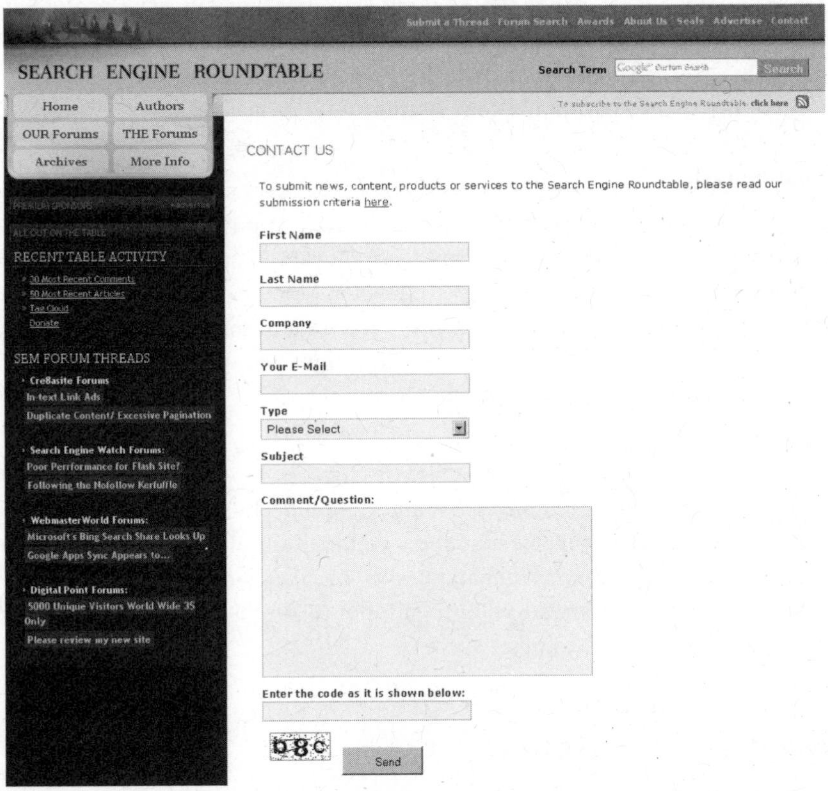

FIGURE 5-12. Keep your contact form basic, but include a CAPTCHA that requires a real person to answer

DEFINITION

A *bot* is a piece of software that traditionally simulates human activity by performing a mundane task automatically or on command. When hundreds of spam comments are posted on your blog in rapid succession, you know that they originated from bots. A CAPTCHA is a distorted image of letters and numbers often used to thwart bots from excessively spamming contact forms and websites.

Keep the comments open

From a social media perspective, the most empowering part of a blog is the ability of community members to be active participants and to share ideas among themselves (and of course, with you). With comments, sharing information is simple, and feedback is almost instantaneous. Whenever you're able, engage in the discussion and continue monitoring it. Also, a word of advice: don't make it difficult for your readers to post comments. Most readers will get turned off if they have to register for an account just to share their two cents. Spam comments can be easily addressed with the installation of a CAPTCHA (or with a variety of known plug-ins, such as Akismet). If you are concerned about offensive comments, you may want to moderate all comments so that you can approve them before they are displayed. Make it easy, not difficult, for the community to take part in the conversation.

Hold regular contests

A great way to build an audience is to hold a contest. Ideally, you should host the contest when you have already built some momentum and have a loyal following. Don't run the contest when there are too few users interacting with the blog, since you might be disappointed by the low amount of interest. Offer prizes, such as a free year of service or a product; in the case of an informational blog, you may be able to solicit prizes from sponsors. People love freebies, and contests are a great way to keep them hooked.

Run polls and surveys

Engage your readers beyond traditional Q&A and comments by holding polls related to your topic. A number of tools are available that allow you to host polls and surveys, such as PollDaddy (*http://www.polldaddy.com*) and Survey Monkey (*http://www.surveymonkey .com*). In one post, you can ask readers for their input (Figure 5-13); in a follow-up post, you can report on the findings of the poll or survey.

How Blogs Are Discovered

With the number of blogs in the hundreds of millions, how does someone find your newly launched blog? One of the most important ways to make sure readers find you is to become visible on related sites within your niche. Read other blog posts and participate in a discussion by leaving a valuable comment. Provide a link back to your blog where necessary. Naturally, don't use this as grounds to self-promote; just be genuine and contribute to the discussion. Over time, if you show a genuine and continued interest in the blogs you participate in, you will likely be able to establish a relationship with the blogger and may be highlighted for inclusion in the site's blogroll or even in a guest post.

> ### DEFINITION
> What is a *blogroll*? A blogroll refers to a list of related sites that appears on the sidebar or footer of a blog. Blogrolls are explicit endorsements by the blog owner of content that is approved for reading.

Before you promote your blog heavily, there is one catch: make sure you have adequate content on your blog before you start engaging in this type of networking strategy. Again, you need to show that you're a serious blogger and plan to provide consistent updates. It is best to wait two or three months before engaging in heavy blog promotion; having a significant

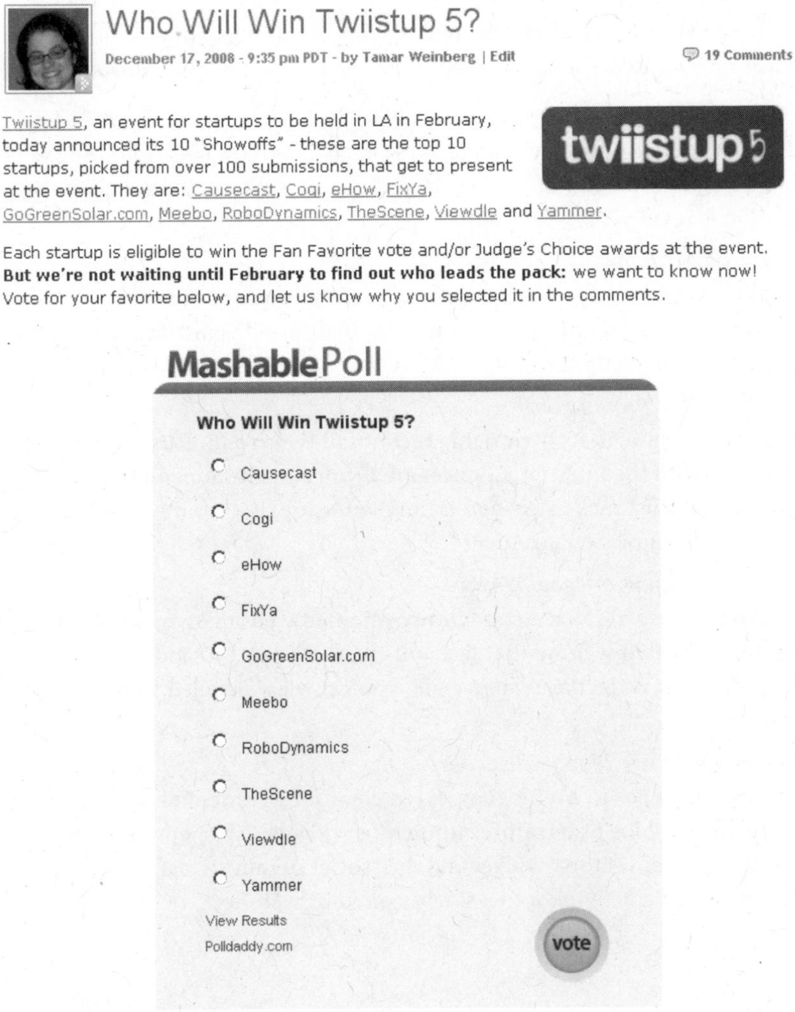

Who Will Win Twiistup 5?

December 17, 2008 - 9:35 pm PDT - by Tamar Weinberg | Edit   💬 19 Comments

Twiistup 5, an event for startups to be held in LA in February, today announced its 10 "Showoffs" - these are the top 10 startups, picked from over 100 submissions, that get to present at the event. They are: Causecast, Cogi, eHow, FixYa, GoGreenSolar.com, Meebo, RoboDynamics, TheScene, Viewdle and Yammer.

Each startup is eligible to win the Fan Favorite vote and/or Judge's Choice awards at the event. **But we're not waiting until February to find out who leads the pack:** we want to know now! Vote for your favorite below, and let us know why you selected it in the comments.

MashablePoll

Who Will Win Twiistup 5?

- ○ Causecast
- ○ Cogi
- ○ eHow
- ○ FixYa
- ○ GoGreenSolar.com
- ○ Meebo
- ○ RoboDynamics
- ○ TheScene
- ○ Viewdle
- ○ Yammer

View Results
Polldaddy.com

vote

FIGURE 5-13. Involve your audience with a poll

number of blog posts indicates that you are ready and serious about your commitment to blogging and establishing networking connections.

Blog Directories

You can promote your blog by actively adding your site to relevant blog directories. Some of these services may be free; others are paid. Here is a brief list of blog directories:

Best of the Web (http://www.botw.org)
> The Best of the Web directory (Figure 5-14) is a paid service ($49.95 one-time review fee), but it is recognized as one of the best paid directories on the Internet, because each submission is reviewed by paid staff.

DMOZ (http://www.dmoz.org)
> DMOZ, also known as the Open Directory Project, is one of the premier free directories and is weighted heavily by Google and other high-profile partner websites. Volunteer editors review the content.

Technorati (http://technorati.com)
> A blog search engine in its own right, Technorati is also a directory for just about every blog that exists on the Internet. Chances are if you already launched your blog, Technorati knows about it. Just make sure you claim ownership of it for more detailed statistics and interactive community engagement.

MyBlogLog (http://www.mybloglog.com)
> MyBlogLog (Figure 5-15) is a community-oriented website owned by Yahoo! and is a community-oriented website that lets you create a widget to add to your website. With installation of the widget or related code, you can view detailed statistics about your site and its visitors.

BlogCatalog (http://www.blogcatalog.com)
> BlogCatalog is similar to MyBlogLog: it is a community-oriented website and blog directory. BlogCatalog also features numerous widgets. Two popular BlogCatalog widgets include the "recent visitors" widget and the "social stream" widget, which integrates your recent social media activity onto a sidebar on your website.

FIGURE 5-14. The Best of the Web blog directory lists blogs in numerous categories, from computers to science to health

FIGURE 5-15. Adding your community to MyBlogLog is as simple as registering your account on the site and then listing your blog under My Sites and Services

These are only a few of many blog directories. There are multiple opportunities for blog promotion via these directories, and engagement on these sites is often a good way to get noticed.

Blog Carnivals

Another way to promote your content is to participate in *blog carnivals* (Figure 5-16). Blog carnivals are community-oriented blog posts that revolve around particular themes. In a blog carnival, one blogger collects multiple links to blog posts about a specific topic. While probably not ideal for corporate blogs, blog carnivals provide a terrific way to gain exposure and build a network. Inclusion in a blog carnival is simple: check out *http://www.blogcarnival.com* and submit your blog post to the relevant carnival. In many cases, you may be able to submit your blog posts to multiple carnivals.

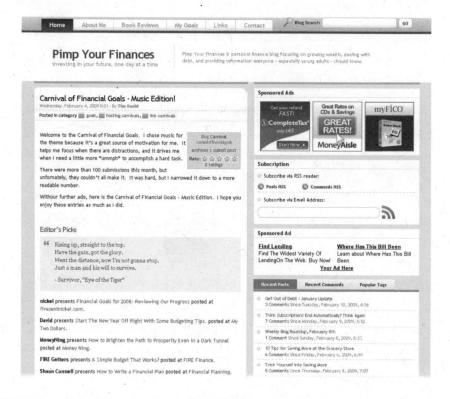

FIGURE 5-16. Blog carnivals aggregate posts that are submitted by multiple users who blog on the same subject matter

Memes

You can also raise awareness through *memes*. Memes (Figure 5-17) are typically a chain of posts that have an original source. The idea behind a blog meme is to start off sharing information about yourself, then tag a number of bloggers and ask them how they would answer the same question. In hundreds of follow-up posts, people start spreading ideas and original links. If you participate in a meme, contact the blogger you tagged to let him know that he's "it," and he'll pay it forward by tagging his friends (and linking back to you).

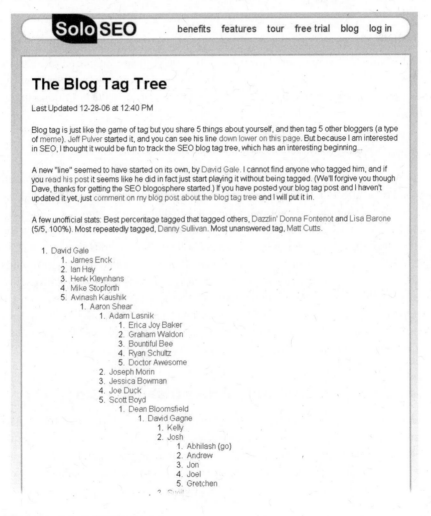

FIGURE 5-17. A manually tracked blog meme

Popular memes in the past have included "5 Things You May Not Know About Me," "My Favorite Charity," and "What Magazines Do You Read?"

TIP

If you engage in blog memes, it is best to tag individuals who occasionally blog about themselves. Some bloggers will not create personal posts on principle, but others are very liberal with the kinds of posts they choose to publish. Many bigger bloggers have to satisfy a tremendous loyal audience and may not necessarily respond to the blog meme because of editorial concerns and fear that they are ostracizing users (since their emphasis is on news items and not personal tidbits). As a common courtesy, if you get tagged in a blog post and cannot participate, let the person who tagged you know so that he can tag someone else.

Social Media

Social media is another way to build awareness. Having your site submitted to a variety of social news sites or social bookmarking sites—and seeing that the content is well received—is a great way to boost exposure. We will discuss more about social media promotion in the latter half of this book.

Writing Projects

Writing projects are yet another way to build awareness. You can leverage your community and get good exposure by asking your readers to offer a bit of advice on their own blogs about a specific topic and then provide you with the published blog post. After the project period expires, link back to all the post submissions. A successful example of this is the Daily Blog Tips Blog Writing Project,[25] where editor Daniel Scocco solicited blog entries related to blog tips and received 122 distinct pointers in a period of 8 days. He then listed all participants' blog posts in a final article on his site and categorized them into sections for easy readability. This community blogging project brought targeted traffic back to his blog, awareness of the project on each individual participant's blog, and a link back to each participant's blog from the writing project hub as a gesture of "thanks" for participation. The relationship is, therefore, beneficial to both the community project hosts and to those who take part in the project.

What to Do If Your Corporate Policy Disallows Blogging

In some corporate environments, legal red tape will prohibit you from engaging in a blog conversation on behalf of your company, or even from starting your own corporate blog. In fact, while blogs are great communication and customer loyalty tools, not every company will benefit from having one. In these instances, alternative methods are available to raise awareness through others' blogs and through the online community.

Get Other Bloggers to Talk About You

A proven tactic time and time again is not to start your own blog, but to get other bloggers who may have a loyal community of readers and listeners to talk about you. Finding these bloggers is not a difficult task. Use Google Alerts to subscribe to alerts for blogs that may talk about your topic area, or use Technorati to search deep in the archives of blogs that may have covered the topic matter in depth. Additionally, once you find these blogs, check each individual site's blogroll, if available. This is another way to find even more relevant blogs. You can then target them with a related press release, product announcement, or even a sample item for the blogger to review.

Blogger outreach doesn't come without a catch, however. As a prolific blogger myself, I've received hundreds of thousands of pitches across the several blogs I am actively involved in that serve to inform me of new product releases and services. I ignore almost 99.9% of these pitches. Today, the best blogger pitch is personalized, brief, and to the point. The traditional press release is too lengthy for most bloggers to read and consume (and is often not well targeted for the particular website). An ideal approach is to introduce the product or service with a short (two- or three-sentence) to capture the blogger's attention and attach related documentation, such as a press release. If a blogger chooses to consider your proposal, he or she will then want to peruse the related material.

Be advised that there are bloggers who will not be open to any communication at all, especially if you attempt to contact them by obtaining their private and personal email addresses (versus contacting them via the public contact forms on their websites), are not targeting the message appropriately, or use excessive spam methods to get them to listen to you. In mid-2008, several eminent bloggers launched a wiki that included the names of PR spammers they had opted to blacklist from future mailings (*http://www.prspammers.pbwiki.com/FrontPage*). Press release firms should acknowledge that any misstep could cost valuable business relationships. Bloggers do not abide by the same types of communication rules as traditional media, and a violation of their freedoms can turn into a public relations firestorm,[26] where the actions of the erroneous public relations professional can be highlighted for all to see. As such, research and due diligence is requested on behalf of bloggers, and polite communication is encouraged.

BLOGGER OUTREACH CASE STUDY: OH! NUTS

Oh! Nuts is a small company based in Brooklyn, NY that specializes in orders for bulk candy and nuts. Faced with a competitive market, Oh! Nuts decided to engage in blogger outreach to alert consumers to its product offerings. By employing numerous blogger outreach tactics, the company successfully gained new customers who previously knew nothing of it, and with a flow of regular blog posts, Oh! Nuts has also seen heightened brand awareness.

Sam Feferkorn of Oh! Nuts explains that he was alerted to the potential of social media and began using blogs as an outreach tool in December 2007 for the Hanukkah holiday. As part of a trial run, the company targeted Jewish blogs. It located Jewish blog directories and reached out to every blogger, offering them a 10% discount and $25 gift. Oh! Nuts received significant interest from these initial outreach efforts, which raised awareness of the company through new blog posts. The mentions of the company and its products in these particular blog posts also improved search engine optimization of Oh! Nuts.

The first successes inspired the small business to do more. Oh! Nuts then reached out to food bloggers and offered them free samples of dried fruits and nuts, with a request that these ingredients be used in a future recipe that the blogger would publish on his/her blog. When making the request, the company never demanded a link, but due to its good faith and genuine outreach efforts, most bloggers ended up linking back to it and some even linked directly to the product page, thereby increasing opportunities for search engine optimization. Feferkorn notes that big bloggers aren't as receptive to this kind of outreach, but the smaller bloggers ("college bloggers," as he calls them) are very appreciative of the gesture.

Feferkorn says that once a blogger publishes a post, his job is not over. Oh! Nuts tries to follow up with everyone it can contact. By looking at who comments on each post, it is able to offer more products to more readers and bloggers. The community, in essence, benefits from being engaged.

The company also finds potential blogs by using Google Alerts for certain keywords and by monitoring the conversation on Twitter.

Feferkorn is satisfied with the work Oh! Nuts put into social media and blogger outreach. He says, "Anytime we see our work resulting in huge sales increases, we can't be happier. We feel that the human element of blogger outreach will play a big part in the way users decide on online shopping."

Develop a Corporate Social Media Strategy

When it's not possible to host a two-way conversation on a blog, it is still possible to engage in the social media community at large. The important thing, however, is to be genuine and transparent about your relationships. Companies that do not blog regularly often consider other alternatives for engagement, and a corporate social media strategy is a good direction to take so that all company employees understand the policies and procedures regarding what is appropriate and not appropriate in social media channels from the company perspective.

Further, a corporate social media strategy is especially important when nobody in the company is directly involved in social media efforts. In essence, the existence of a corporate media strategy encourages employees to involve themselves in social media matters without fear of company crackdowns that may silence ardent company supporters who love to talk about where they work.

In a corporate social media strategy, it is important to lay down the law on what employees can and cannot do according to the internal corporate policies and objectives. If the legal department has restrictions on how company representatives communicate with the public, this should be accounted for in the social media strategy.

An ideal social media strategy considers multiple "dos and don'ts" for how an employee should conduct himself on a variety of social media channels. For example, in alignment with the mission of this book, it may be necessary to write a strategy that discourages the direct hard sell of any particular product or spam submissions on social news sites. Perhaps it is best to begin with an introduction to social networks and to simply encourage staff members to join these sites, be social within them, and to consider communicating wisely with other community members. If social media involvement is encouraged among staff members, it may be ideal for them to join the sites without actually identifying themselves at all with any particular company or brand; at first, they should simply learn how to use the websites, how to network, and how to benefit from the sites in a way that will encourage long-term enjoyment. In some corporate policy guides, promotion on behalf of the business is strictly discouraged until employees get a handle for the social site; only after a designated number of weeks or months is the employee able to identify with the company (and to promote on behalf of the company in some instances).

Of course, the company policy must account for something far more important in the long run: what you say can hurt you. Professionalism is vital. Emotions should never interfere with logic. Company employees should never say something without absolute certainty of its truth. Writing a blog comment that is "off the record" does not fly with a corporate social media strategy, because search engines are always watching, or the people you converse with may make your commentary public. Therefore, it is imperative that you be extra careful when engaging on social sites and with individuals you may or may not know. When in doubt, ensure that a qualified company spokesperson speaks on the company's behalf. Because just about everything can be picked up by search engines, you must tread extra carefully, especially if the communication or communicator is related to a particular corporate entity.

In any social network, it should be clearly stated somewhere in the user's profile that opinions are those of employees and not of the company. This disclosure prevents the perception that the employee is working on behalf of the company to communicate its goals at all times. Moreover, to further distance themselves from the company, employees should not use corporate imagery or proprietary information on their own social media accounts.

Corporate Social Media Strategy for Blogging

The corporate social media strategy should also address employees who prefer to maintain their own personal blogs. Outside the workspace, employees should be allowed to keep blogs that are absolutely not associated with the company, especially if the company does not wish to participate in blogging or maintain its own blog.

Several considerations should be made for those who blog and are also associated with a company. A good example of a well-known blogger in the search engine industry who maintains his personal blog is Matt Cutts of Google. He uses the blog to talk about his personal affairs and adventures in technology, but he makes a clear distinction that he is not communicating on behalf of his employer in his blog's disclaimer:[27]

> I'm one of several Googlers who answer questions online and sometimes for the press. I usually handle questions about webmasters or SEO, so in those areas I'm more likely to make sense and less likely to say something stupid. If I post something here that you find helpful as you build or manage your web presence, that's wonderful. But when push comes to shove: This is my personal blog. The views expressed on these pages are mine alone and not those of my employer.

In most corporate social media strategies for bloggers, employees are allowed to mention who they work for and share generic information about how much they may love their jobs. That's the extent of their affiliation with the company. Naturally, common sense is expected, and discussing proprietary technology, private company policies, and super-stealth projects is completely off limits.

A social media strategy should specify that company employees state in their blogs that they are speaking on their own behalf, and not the company's, as demonstrated in the example of Googler Matt Cutts. For example, the footer of the blog (and/or the "About" page) should say something like, "The views expressed within this blog are my own and do not necessarily represent the positions or opinions of [my company]."

In some more stringent company policies, bloggers may be held liable for their comments. Therefore, the social media strategy documentation may need to include a note for bloggers to maintain caution when they write anything that may be perceived to be libelous, defamatory, obscene, or proprietary. It's always important for companies to reinforce the fact that bloggers can be held accountable for any missteps, and that consequences of their actions could reflect poorly on the employer.

In some instances, the media may contact individuals associated with a company for official comment on its behalf. There should be a clearly defined company media contact for any and all such inquiries, and bloggers should be discouraged from speaking directly to the press.

Other best practice guidelines include honoring copyright law when images and text are referenced, showing respect toward peers, ensuring that there is context for all arguments, and encouraging private feedback (and not making these communications public in the interest of professionalism).

Summary

Blogging is one of the oldest and most successful approaches to social media, as there are hundreds of thousands of content creators and consumers. By themselves, blogs can be very influential, but sometimes this influence can be dangerous, as evidenced by the loss of $4 billion in Apple stock, which resulted from inaccurate blog reporting. This example, however, also illustrates the sheer power of blogs and how they are penetrating the online space. Today, bloggers are newsmakers, but the blog also empowers individuals to share information about their personal experiences. For business purposes, blogs are a great way to connect with an audience and to attract customers and individuals already loyal to your brand.

Blogs can be consumed in a variety of ways, from visiting the home page directly to subscribing to email alerts. Feed readers are popular among tech-savvy individuals, while others opt for personalized home pages such as My Yahoo! and iGoogle. With blogs' RSS capabilities, it is easy to consume blog content from just about anywhere.

Blogging platforms are a lot more powerful than static web pages for a variety of reasons. They often come equipped with a WYSIWYG editor, which means that bloggers do not need to possess extensive HTML skills in order to create a professional-looking blog post. Moreover, blog software includes built-in RSS functionality so that readers can consume blogs on their chosen platforms, be they mobile phones or customized start pages. Finally, blogs can ping search engines and blog engines, and this usually results in rapid indexing of blog content in search results.

In this chapter, we reviewed five popular blogging platforms, including the very popular WordPress application suite, which boasts millions of users. MovableType and TypePad are also widely used blogging platforms. Google's Blogspot (Blogger.com) is extremely popular because it is easy to set up. ExpressionEngine is the least popular of the bunch, but it has some powerful features not offered by any of the aforementioned platforms. All of these tools have different pros and cons, and you will need to choose your platform based on whether you want to pay for hosting or design the blog yourself, and whether you have the technical know-how to address problems, among other considerations.

Once your blogging platform is set up, you should ensure that the tone of the blog is genuine and that you are not merely regurgitating corporate lingo. The Southwest Airlines blog has been extremely successful because its bloggers share real-life experiences that humanize the company. Other blogging factors pertain to the style of the blog and not just the voice. Adding visual elements and eye candy is a social media strategy that has fared well, but you should consider good clean text and a reasonable strategy of linking to related bloggers and articles.

In blogging, the standard three-paragraph release isn't necessarily the best approach. Powerful headlines that draw attention are crucial, but in terms of the body of the article, you can consider strategies such as compiling lists, writing in-depth how-to articles, telling stories, interviewing experts, reviewing products or related services, and offering regular features that readers have come to expect from your online publication.

Inspiration for blogging topics can be found in email alerts, related blogs, social bookmarking sites, news, and via popular trends. When in doubt, do a search and you'll likely come up with a lot of ideas that are relevant to your readers.

Enhancing your blog is easier than ever with widgets such as CloudShout. You can also dress up posts with charts and video. There are some extensive commenting solutions available, such as Disqus, a community that supports nested comments, while Seesmic lets you actually submit recorded video comments. All of these solutions give blogs a more interactive feel.

If you have blogworthy content and you're attempting to build an audience, it is time to start involving your readers in the conversation. Of course, by default, you'll likely allow them to comment on the blog, but you can also invite questions via a contact form, which you will need to set up yourself. Fortunately, many blog applications have plug-ins to make adding a comment form to a blog template a relatively simple task. It is also possible to invite your readers to share commentary via "ask the readers" columns, contests, and polls and surveys.

Discovery of blogs can be difficult at first, but one way to get noticed is to make your existence known. Participation on blogs in your industry and niche is one of the most useful ways to get your name out there. The right networking skills can translate into long-term relationships and perhaps even translate into blogroll inclusion, which is essentially an endorsement from another blogger who admires your content. Other promotional strategies include participating on blog carnivals, submitting your blog to directories, hosting or participating in a meme, engaging in social media promotion via social news sites or social bookmarking sites, and by running a group writing project.

While all these strategies may sound great on paper, some companies just will not allow their employees to blog on their behalf. Some companies also won't give you the green light to start your own company-oriented blog. However, community participation can still yield brand awareness. Blogger outreach is a strategy that also works; instead of creating your own blog, contact bloggers who will talk about your product. If your offer is compelling, the blogger will likely be interested.

Having a corporate social media strategy is also a smart option for companies. This strategy should accommodate best practices for community engagement and blogging, and ensure that anyone involved in these social channels is doing so on his/her own and not on behalf of the company.

Endnotes

1. *http://www.online.wsj.com/article/SB118436667045766268.html*

2. *http://www.technorati.com/blogging/state-of-the-blogosphere*

3. *http://www.sifry.com/alerts/archives/000493.html*

4. *http://www.businessweek.com/bwdaily/dnflash/content/feb2008/db20080219_908252.htm*

5. *http://www.technorati.com/blogging/state-of-the-blogosphere*

6. *http://www.pewinternet.org/PPF/r/144/report_display.asp*

7. *http://www.pewinternet.org/PPF/r/113/report_display.asp*

8. *http://www.pewinternet.org/PPF/r/186/report_display.asp*

9. *http://www.pewinternet.org/PPF/r/240/report_display.asp*

10. *http://www.pewinternet.org/PPF/r/240/report_display.asp*

11. *http://www.techipedia.com/2007/social-media-impacts-journalism*

12. *http://www.pewresearch.org/pubs/1066/internet-overtakes-newspapers-as-news-source*

13. *http://www.emarketer.com/Article.aspx?id=1006146*

14. *http://www.technorati.com/blogging/state-of-the-blogosphere/brands-enter-the-blogosphere*

15. *http://www.engadget.com/2007/05/16/iphone-delayed-until-october-leopard-delayed-again-until
    -januar*

16. *http://www.news.cnet.com/8301-10784_3-9719952-7.html*

17. *http://wordpress.org/extend/plugins*

18. *http://wordpress.org/extend/themes*

19. *http://www.typepad.com/pricing/index2.html*

20. *http://www.socialtext.net/bizblogs/index.cgi*

21. *http://www.prnewsonline.com/awards/platinumpr/event_info.html#Blog*

22. *http://www.readwriteweb.com/archives/the_stats_are_in_youre_just_skimming_this_article.php*

23. *http://www.problogger.net/archives/2006/11/27/dont-be-an-insular-blogger*

24. *http://www.copyblogger.com/the-5-immutable-laws-of-persuasive-blogging*

25. *http://www.dailyblogtips.com/blog-writing-project-tips-tricks-final-list*

26. *http://www.techcrunch.com/2008/12/18/meet-lois-whitman-the-poster-child-for-everything
    -wrong-with-pr*

27. *http://www.mattcutts.com/blog/disclaimer*

CHAPTER SIX

Microblogging Magic: How Twitter Can Transform Your Business

Twitter is a free microblogging service that allows users to communicate with one another using short text-based messages that can be a maximum of 140 characters in length. Twitter provides robust tools for sending and receiving updates on a variety of devices and through a plethora of tools. Launched in 2006, today Twitter has over 14 million accounts—with much of the growth attributed to celebrity adoption in the first quarter of 2009—and serves both professional and personal needs. Companies are using Twitter to tap into business prospects, influencers, and customers. As such, although it relates to topics discussed elsewhere in this book, Twitter deserves a chapter in itself.

A History of Twitter

The original goal of Twitter was to have users answer, "What are you doing?" in 140 characters or less. When the service launched in 2006, that is exactly what people were sharing. Users of the service provided updates about the food that they ate for dinner, the places they went, and the people they met. Initially perceived by most individuals as a tool that served no purpose and was a waste of time, Twitter's original influencers realized that there was more to the service than just users talking about nonsensical happenings in their everyday lives. Twitter's capability of connecting individuals seamlessly through so many devices created a sense of intimacy and closeness and forged incredible bonds. Individuals not only decided to answer the question, "What are you doing?" but also used Twitter to share their thoughts and feelings, and to let the world know exactly what was most important to them.

In mid-2007, Twitter exploded at the SXSW (South by Southwest) conference, as it enabled conference participants to keep up with the multitude of sessions and also to schedule in-person meetings. After Twitter made its debut on the big screen at the conference, with conference organizers encouraging participants to sign up for updates, more and more people started using Twitter and a community blossomed. Twitter became the tool of choice over more invasive tools such as Instant Messenger, because people could tune into it at will, it had social elements, it appeared to be scalable, and through its coding API, it boasted a variety of plug-ins and add-ons (we will discuss some of these plug-ins and add-ons later on in this chapter).

The SXSW conference showed that a multitude of individuals with similar interests could congregate on a single site to connect and communicate with new and old contacts. Once the conference wrapped up, users and businesses started tapping into Twitter and taking advantage of its offerings. Thought leaders began to share their insights with their peers and began participating in ongoing valuable discussions by publicizing related news articles and blog posts. Marketers began to see the value of connecting with individuals in their industries to share and to discuss their views. Businesses realized the value in getting immediate feedback about their products and brands; they noted that Twitter was a remarkable tool for reaching prospective customers and offering quick and painless customer service. At the same time, those businesses realized that they could also promote new services and products to a targeted audience. Twitter has become increasingly valuable for connecting consumers with companies for rapid assistance.

Twitter for Business Is Born

Once people discovered that Twitter had a real purpose in the business world, the service became saturated with businesspeople and marketers alike. These groups have taken advantage of the ability to tap into the targeted audience for relationship-building, networking opportunities both online and offline, achieving business objectives, and personal gain. Since Twitter is a distributed communications system, messages can be sent from nearly any device that has Internet connectivity. Further, since Twitter is a broadcasting medium, if you have a sufficient number of friends on Twitter (followers are people who subscribe to your updates), you can share a single 140-character message with hundreds or thousands of people. And finally, since people love to share high-quality *tweets* (messages that are sent via Twitter) with their followers, if you provide valuable content, people will likely *retweet* it (share it with their followers), thereby letting your content travel even farther than previously imagined.

But how does this factor into Twitter for business? Simply stated, Twitter revolves around personal relationships, and the most successful Twitter users are those who engage in facilitating and building friendships. Some of the most successful business entities are those that actively monitor the conversation, show concern for the well-being of their customers, and provide in-depth and quick customer service. Of course, while this may seem like a business attempt at addressing reputation management issues, the fact that quick customer service is only a tweet away has driven many individuals to use Twitter, because it bypasses slower and more traditional means of contacting support representatives.

In reality, Twitter has proven itself to be a reputation management tool, but it is also a tool that brings people all over the world closer together for business and personal goals. Businesses realize that consumers are talking about them, and they are taking advantage of the opportunity to respond quickly. You don't need to hire market research companies anymore.

TWITTER FOR CRISIS COMMUNICATIONS

As a broadcast medium, Twitter is more than just a business tool. It has served as a source for breaking news. One event in recent history that captures the essence of Twitter as a crisis communications tool is the Mumbai terrorist attack of November 2008. Firsthand reports of the terrorist attacks occurred not on traditional news outlets, but on Twitter. Approximately 80 tweets were sent every 5 seconds with eyewitness accounts of the attacks. Some of these tweets were reported from inside the hotels that had been taken under terrorist control. Others came from volunteers who tried to organize blood donations and provide other public services.

What did Twitter's involvement in Mumbai prove? It showed that news can break first from nonjournalists. Furthermore, cell phones are incredibly powerful for communication, especially when tapping into a broadcast medium such as Twitter for maximum exposure. Indeed, it was through individuals armed with mobile devices that the attacks on Mumbai were made so widely known; without these tools, the Mumbai attacks would have been a nonevent in the technology community.

Using Twitter

Twitter is easy to use, and even easier to sign up for. Once you're set with your account, customize your profile. Add an avatar, personalize your background with a unique picture, and write a brief biography about yourself. Then start following people. Typically, you will not know who to follow, but hopefully you'll know a person or two who is already using the service.

If you don't know who to follow, you can search for existing users via the Find People link at the top of the page. Use this to locate existing individuals on Twitter, invite users from other predefined email networks, or invite users via alternative email addresses. Until you start following individuals, your timeline will be empty. Figure 6-1 shows a relatively active Twitter profile: my own.

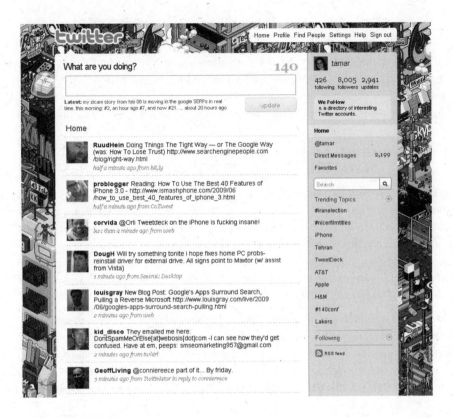

FIGURE 6-1. A sample Twitter home page

If you want to find other followers, take advantage of Twitter's amazing search tool, available at *http://search.twitter.com*. To find followers using this method, search for terms that you have an interest in and follow users based on their involvement. Chances are, you'll find a lot of users with similar hobbies and business affiliations. Take advantage of any and all connections, especially if you're starting out.

Of course, you're only following users at this point, but how are they going to follow you? The best way to do this is to get involved. Look at what other people are tweeting and then share your thoughts. You can write a simple response to a comment using the @ designation, so for example, if you are responding to me on Twitter, you would preface any response with @tamar and then proceed with your message. Believe me, people will take note. This activity shows up on the Twitter replies page (@*yourname* on the Twitter sidebar, as shown in Figure 6-1). Further, via Twitter Search, you can always subscribe to feeds with specific search strings so that you can keep abreast of what people are saying about you.

How to Achieve Business Goals Through Twitter

Twitter is a relevant tool for businesses to tap into a vast audience of users. After all, every member of Twitter is a consumer and most, if not all, of your users on Twitter are already using the tool to talk about you. For businesses, Twitter is exceptional at connecting users with the companies they are already interested in. Twitter, therefore, can generate sales, act as a customer service tool, promote brand awareness, and even capture new prospects as clients. In the following examples, you will see how companies already using Twitter have found success.

Using Twitter to Generate Sales: The Big Guns

Can you make money through Twitter? If you're PC manufacturing company Dell, you certainly can. Over a period of approximately 24 months, Dell used Twitter to alert its users to exclusive deals and made over $3 million in revenue that can be sourced directly to the microblogging website.

While $3 million isn't much for a company like Dell, it does show that companies *can* make this kind of revenue using Twitter, and it's not hard to achieve. Dell has a significant presence on Twitter. News, community sites, offers and promotions, and international blogs associated with the Dell brand name are all active on Twitter. Further, numerous representatives from Dell's different departments, such as its corporate communications team and sales team, are also actively present on Twitter (*http://www.dell.com/twitter*).

Since you can easily subscribe to updates from a variety of vendors through Twitter and access the site as you please, it's relatively easy for users to tune into updates and act upon them. That's exactly what Dell did, and that's exactly how it made seven figures in 18 months.

If you're looking to Twitter as a source of income, consider a strategy that allows you to broadcast sales that you can offer your Twitter followers exclusively. Perhaps you want to offer a deal that is not available anywhere else; use Twitter to share a coupon code that will be associated only with the Twitter account that you are posting from. Alternatively, create a custom URL for a Twitter-specific deal and promote that deal only on Twitter and not through other advertising channels. At least once a week, post a great deal that people will want to talk about. This is how you leverage word-of-mouth marketing via retweeting. To ensure that your customers know about your presence on Twitter, create a page for your website advertising your Twitter account or send an update to your customers via your standard newsletter.

Just remember that this strategy worked for Dell's Home Outlet Store (*http://www.twitter .com/delloutlet*).

Using Twitter to Generate Sales: The Smaller Companies

You may feel that Twitter really works best for the larger companies. After all, more people know about Dell products than your small business. However, small businesses have also successfully navigated the Twitter landscape and leveraged the community to see success.

Namecheap is a domain name registrar that capitalized on Twitter's community in two contests held in late 2008 and early 2009. As an active user on the service, Michelle Greer, the company's marketing specialist, noticed that Twitter can be a successful driver of traffic and sales without a significant financial investment. As such, the company ran a trivia promotion in which it asked questions every hour over the course of several weeks. The first three people to answer correctly were given a $9.69 credit in their Namecheap accounts, the price of one domain name, and the individuals who had the most answers correct at the very end of the contest received iPods.

The contests were extremely successful and saw thousands of individuals scrambling to participate and win prizes. However, while the community gained from this, Namecheap really emerged as the winner of the contest. During the December 2008 contest, the company saw an over 2000% increase of Twitter followers, a 20% increase in new domain name registrations, and 139 backlinks pointing to the specific contest page on the namecheap.com domain in addition to countless others pointing to the home page.

At the end of the day, this strategy was a significant time investment, and Greer acknowledges that a Twitter campaign of this sort takes a bit of work. After all, more than 600 questions were written for the purpose of this trivia contest, and $17,000 in domain names were offered as prizes. Greer says that four people participated in maintaining the account during the contest's run: she wrote all of the questions and monitored the answers herself; the CEO of Namecheap, Richard Kirkendall, also occasionally monitored the account to ensure the smooth running of the contest; the CTO of Namecheap, Mohan Vettaikaran, wrote a script compatible with Twitter's API that fed the questions to Twitter in a systematic fashion; and a fourth employee of the company fed the questions into the tool that was written by the company's CTO.

Despite the manpower and commitment to the contest, Greer says Twitter is an extremely affordable alternative to other available solutions: "Honestly, it helps Namecheap become a better company because we can get feedback directly from our customers at a much lower expense than using surveys or consultants. We offer value to our customers with free domains, and they help us improve our company. It is a win/win for everyone."

Using Twitter for Customer Service

In social media, regardless of whether you are going to engage or not, conversations about you will be ongoing. That's clearly the case with Twitter. If you are part of an established business, a Twitter search on your company will likely produce hundreds and perhaps thousands of results. You'll get almost immediate feedback from customers and find out exactly what they think about your service or product offerings. Naturally, you'll likely want to respond on behalf of your company to some of the not-so-positive commentary. As such, Twitter has become a grounds overflowing with customer service inquiries, and representatives of companies have taken the steering wheel on the situation to alleviate customer concerns.

The Comcast story

Comcast is one example of Twitter customer service being taken to the next level (Figure 6-2). Frank Eliason, Director of Digital Care at Comcast, has successfully managed to set a customer service example for Twitter. According to Eliason, Comcast has been monitoring the conversation for years. However, over a year ago, Twitter was referred to the company as another potential communications channel. Initially, Comcast was hesitant to get involved on such foreign ground, but after interactions with bloggers and customers who were engaged in the technology space, the company realized that there was much benefit to be had by reaching out to these customers and having conversations with them. Eventually, Comcast's involvement earned it mentions in blog articles and mainstream media, and because of this newfound fame, other companies are striving to emulate its example.

FIGURE 6-2. ComcastCares on Twitter

After several months, Comcast employees also saw the value of the community on Twitter. After all, the community considers them not only representatives of a company, but participants of the community. On Twitter, your affiliation with your brand perhaps reinforces real relationships with people.

In terms of monitoring, the company does not use any extravagant tools to search for mentions of its name. For Comcast, Twitter Search is the tool of choice. There are four Comcast employees manning Twitter at any given time.

Comcast customers like Rebecca Kelley are pleased with Comcast's active involvement on the social network.[1] Kelley writes of her engagement with ComcastBill on Twitter, "He remained prompt, pleasant, helpful, and understanding, which all equated to a positive customer experience for me." Further, after her experience with Comcast ended, she "was extremely impressed with how Comcast leveraged Twitter to reach out to its customers and efficiently manage their reputation."

Eliason is happy with how Comcast is engaging with the community via Twitter as well. He says, "I think it has been a tremendous way to communicate with our Customers and obtain Customer feedback quickly and easily. As we strive to improve Customer Service, this is a great way to obtain the Customer story in their own words. It is a great way to help improve the overall experience."

The JetBlue story

JetBlue is another company that has been using Twitter successfully (Figure 6-3). Consumer issues blog Consumerist.com highlights[2] an event that illustrates just how effective JetBlue support via Twitter has been. In the example, a Twitter user merely said, "JetBlue, I need a wheelchair!" on her Twitter stream. Shortly thereafter, she attempted to call JetBlue customer service to achieve this goal. But before she even spoke with an agent, JetBlue on Twitter (*http://www.twitter.com/jetblue*) came to her aid and connected her with someone who was able to offer her immediate assistance.

Morgan Johnston, a manager for Corporate Communications at JetBlue, explains that he uses new media tools to improve the company communications. Before JetBlue actually engaged heavily on Twitter, he observed the community and listened to see how individuals reacted and how other users responded. Initially, the company merely scanned Twitter for mentions of its name. Later, after noting some misconceptions about the company, JetBlue realized the benefits of becoming actively involved. This has given JetBlue the edge of being more than just a "faceless company."

Johnston prefers TweetDeck, an Adobe AIR–powered application, to monitor mentions of JetBlue on Twitter. He says that this helps him run several search queries while monitoring direct messages in a single graphical interface (TweetDeck is discussed later in this chapter).

NOTE
Adobe AIR is a new cross-platform technology that brings lightweight desktop applications to any Windows, Macintosh, or Linux computer.

FIGURE 6-3. JetBlue Airways on Twitter

With name monitoring comes more than just customer service, according to JetBlue. Johnston says, "I liken our current Twitter account style to that of an information booth—but it's so much more." Indeed, JetBlue has worked to communicate new developments in the company, such as its interactive "Where We Jet" map, and also to reach out to customers, often privately. Johnston admits that JetBlue's public Twitter stream features less than 25% of the company's total communications on Twitter and most customer service issues are handled via direct messaging.

The public perception of JetBlue, especially given its engagement on Twitter, has been positive, but Johnston acknowledges that there are some detractors. "There have been a few people who've felt that our @JetBlue account isn't capable of providing the level of customer service they'd like to see. However, for the constraints of the tool, I think we're doing a respectable job of keeping a level of dialogue going, and as always we continue to adapt our strategy according to what our customers would like to see from us on Twitter," he notes. Indeed, given the limitations of Twitter, you'd think customer service wasn't possible, but JetBlue shows that it definitely is. And it's doing a lot more than other companies have even considered.

Using Twitter for Brand Awareness

When you provide value to your followers, you will find that more people will become aware of you. When you are involved heavily in a community, and you bring more and more people along for the ride, your awareness goes up even further. Finally, if you have fun above all else, those around you will feel good and enjoy your company.

The Zappos story

Online shoe retailer Zappos has a unique approach on Twitter. The company is heavily involved on the microblogging site, with over 400 employees using the service.[3] The idea of Zappos' involvement on the site stems from its desire to convey a positive company culture, which CEO Tony Hsieh says translates into great customer service and increased growth of the brand. With Twitter, Zappos is able to successfully form personal connections with customers and prospects just as they'd be doing on the phone.

Hsieh had been using Twitter since SXSW in 2007. When he saw the benefits of the service, he decided to introduce it to the company. In the spring of 2008, Zappos made Twitter a company priority. Today, new employees are introduced to Twitter during staff orientation, and Twitter classes are offered to all employees. Employees are encouraged but not forced to use the service, but as Hsieh says, "When there are so many Zappos employees using it, many employees just gravitate naturally to it."

To monitor the conversation, Zappos developed a tool that scans the site for mentions of the company. This URL is publicly accessible at *http://twitter.zappos.com* (Figure 6-4).

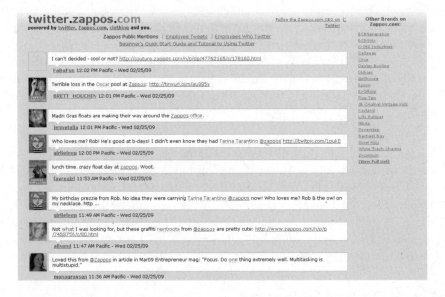

FIGURE 6-4. Zappos' Twitter page

Today, the company culture is obvious on Twitter. Zappos clearly shows it can have a good time, but is also happily engaging the community on Twitter. In a Zappos blog post, Hsieh acknowledges that while reputation management crises are abundant online, the "reverse is true as well. A great experience with a company can be read by millions of people almost instantaneously as well."[4] And that's exactly what is happening with Zappos. Twitter users feel overwhelmingly positive toward the Zappos brand with the company's involvement both online and offline (through networking events, such as a party that corresponded to the PubCon search engine marketing conference in November 2008 in Las Vegas).

Internally, Twitter has connected the company in unforeseen ways as well. Hsieh recounts a story in which an employee tweeted that she wanted a cheeseburger, though she wasn't expecting one. It just so happened that another Zappos employee was on her way to buy one when she saw that tweet. Within 10 minutes, Hsieh says, the employee longing for the food had the cheeseburger on her desk. These kinds of incidents help build the company culture internally.

Hsieh is very pleased with the company's Twitter involvement. He says

> I think [Twitter has] helped our culture in that employees have more of an opportunity to learn about each other personally. Twitter also allows us to expose our company culture to the world. I've been able to form personal connections with many of our employees and customers through Twitter, and it helps everyone feel like they are dealing with real people, not a faceless corporation or CEO.

Indeed, with Zappos' involvement on Twitter, its brand is growing stronger and customers are enthusiastic about their engagement with the company.

Using Twitter for Client Acquisition

Twitter has already been illustrated as a tool to generate sales, to assist customers with problems, and to increase a company's brand awareness. Is it possible that Twitter can also help you locate prospects and clients?

The Verizon story

Julio Ojeda-Zapata writes on the Touchbase Blog[5] about an interesting client acquisition scenario in which one customer was wooed by two competing companies. In this particular instance, a doctor was disappointed with an arrogant Verizon technical service representative and expressed his dissatisfaction on Twitter. Immediately, a representative from Verizon came to his aid.

But Verizon wasn't the only company who noticed Dr. Gary Kerkvliet's disappointment. Frank Eliason of Comcast also observed that Kerkvliet was disappointed with his recent interaction with Verizon customer service. Initially, Eliason merely wanted to help the doctor smooth out his cable problems by offering technical advice, but after it became evident that the Kerkvliets were not interested in continuing their cable service with Verizon, Eliason was on the sidelines helping the family transition to Comcast—on a weekend, no less.

Meanwhile, due to Verizon's involvement on Twitter, Dr. Kerkvliet became a champion of both companies. He has nothing negative to say about Verizon anymore. In fact, he says that if you are troubled by customer service, all you need to do is send out a tweet to get the companies to listen.[6]

How can you acquire customers on Twitter? It's as simple as monitoring for your competitors or industry terms and then participating in the conversation when it seems right. You may not want to make the first overt sell, as this may turn off the potential candidate. Be genuine and offer to help first, just as Eliason did before it became obvious that Dr. Kerkvliet was actually interested in pursuing service with the competitor. Eventually, you may find new and unforeseen opportunities to capitalize on the investment and later, to monetize—and the only thing you need to do is listen.

Using Twitter As an Official Communications Channel

You can broadcast events on your website, but if nobody knows about your website, your message will have limited reach. After attacks in Israel's Gaza region, the Israeli Consulate of New York decided to use Twitter to hold a "citizen's press conference," which it used to articulate the government's stance on the recent response by Israel. For two hours, the consulate took questions that were tagged with "#AskIsrael" and gave honest and genuine answers about how the country perceives and is addressing the terrorist threat.

The press conference was an incredible success. While over 750 questions were asked of the Israeli Consulate in a short period of time, the consulate was able to successfully field 55 of them. Later, the Israeli Consulate's official blog (*http://www.israelpolitik.org*) was used to answer other pressing questions that could not be addressed within the allocated period.

The Israeli Consulate used Twitter because of its vast reach and exposure and because traditional media is not necessarily the tool of choice among those heavily involved in the technology arena. Within 24 hours of the launch of the Israeli Consulate's Twitter account (*http://www.twitter.com/israelconsulate*), the account had over 2,000 followers, and many of these individuals were engaged in dialogue surrounding the press conference and the incidents in Israel (Figure 6-5).

The press conference was never intended to sway opinion, but rather to express Israel's honest views on the recent attacks. As shown in Figure 6-6, the consulate provided detailed and thorough answers to the best of its ability given the 140-character limit. The feedback on the Israeli Consulate's involvement on Twitter, especially because of its government status and its openness and involvement on this social media channel, was overwhelmingly positive and has likely set a precedent for other government entities to do the same in the future.

FIGURE 6-5. The Israeli Consulate on Twitter

Other Business Uses of Twitter

Because Twitter is always in use and feedback is always immediate, it is a perfect tool for more than just the aforementioned business objectives. You can use Twitter to establish thought leadership, get instantaneous feedback about pressing matters, build a valuable network (that may likely spread to other social networks) to connect with business partners, and more.

Build a Personal Brand and Establish Thought Leadership

As an active social platform, Twitter is a great tool to build your personal brand. For example, some social media enthusiasts will share relevant thoughts and articles that they've found and written to their audience, and in turn become well known as social media content producers. The more consistent and relevant your tweets are, the more likely you are to gain support as a leader in your industry. Small business expert Anita Campbell has successfully used Twitter, in addition to her blog, to put herself in the forefront as an individual knowledgeable about topics relating to small business. She says:

> Anyone who isn't using Twitter is missing out on the most powerful back-channel on the Web today. When I first signed up for Twitter I "didn't get it." I was baffled as to why smart business people would waste time writing short messages about being delayed at the airport or what they happened to be reading. But I soon realized that Twitter was where energetic conversations were happening, including with my market of small business owners. There's a crackle of energy on Twitter. If you want to be perceived as a thought leader, today you have to be participating on

Twitter and helping create that energy. Otherwise you could miss out on making connections and expanding your network and with it, the opportunity to expand your influence.

FIGURE 6-6. The Israeli Consulate hard at work addressing the community during the "Citizen's Press Conference" in December of 2008; Consul David Saranga of the Media and Public Affairs Division dictates while four employees research and provide 140-character answers

Get Feedback Instantly

Once you achieve an active and loyal following on Twitter, one of the biggest benefits of using the service is the ability to get people to answer questions quickly. To illustrate this phenomenon, I used Twitter in December 2008 to ask my peers how they defined success for a blog post.[7] Within moments, I had 24 solid, thoughtful, and thorough answers. Putting it simply, people share their thoughts immediately about what is important to them, or they can give you important feedback on pressing matters. Is that color scheme good? What do you think of our new product? This information can give you valuable insights about your company and can provide you with actionable information and inspiration for future internal or external projects.

Network with Like-Minded Individuals

Earlier in this chapter, we examined how you can use Twitter's search tool to find followers. This is also a way to network with people who have similar industry interests as you. Search for keywords related to your interests (for example, consumer affairs, search engine optimization, small business, information technology, Linux, or any combination of the above) and start following the individuals who are tweeting with content that you find interesting. Of course, you don't necessarily have to aggressively search for users; they may also be looking for you. Be sure to keep your Twitter stream of consciousness relevant and interesting so that users know exactly who they are connecting with.

Job-Hunt, Organize Events, and More!

Once you build your personal brand and network successfully on Twitter, it can also connect you with new career opportunities. You can also use Twitter for organizing events and meet-ups. Additionally, Twitter can assist with creating to-do lists or goal setting using a tool like Remember the Milk (*http://www.rememberthemilk.com*). With Twitter's widespread adoption, it is becoming more of a productivity tool. The aforementioned capabilities of the service only scratch the surface.

YAMMER: A PRIVATE TWITTER FOR THE BIZ

With Twitter's triumphs in the technology sector, several clones have emerged, some of which have been successes and others of which have been flops. Yammer is one of the more victorious Twitter clones. Intended for businesses, Yammer requires you to maintain a company email address in order to join. With that address, you can converse with your colleagues in a private room that is available only to those in your company. Since Twitter is public, Yammer is the preferred tool for companies. Yammer confines company updates to a privately controlled channel and, as such, helps avoid unnecessary email overflow. It has performed very well since its launch in mid-2008 and many companies use it faithfully.

Yammer offers free accounts, but administrative functionality is available for $1 per member per month after a three-month trial. This functionality allows you to remove messages and members, set passwords, restrict Yammer access to IP or VPN, brand your page with your company logo, and give other people administrative access.

Further, Yammer is web-based, but also features an application built on the Adobe AIR platform, thereby making it suitable for individual use on Mac OS X or Windows without the need to have a browser open.

Tools of the Twitter Trade

In due time, you'll be a Twitter addict like most users on the service. Eventually, then, you'll likely want to take advantage of Twitter tools to make your life easier. There are hundreds of available tools, and some are more popular than others. In this section, we review some of the more widely known Twitter tools that are in use by the Twitter community. This list is by no means exhaustive; there are hundreds of applications that serve the Twitter community.

Twitter Clients

On your computer, Twitter can be accessed via several different avenues, from the web interface (at *http://twitter.com*) to applications that you can install on your desktop to access Twitter without having the web page open at all times. These applications also boast special features, such as searches and support for additional microblogging applications.

Seesmic Desktop (http://desktop.seesmic.com)
> Seesmic Desktop runs on Adobe AIR and connects users to Twitter so that they don't have to use the web interface. Seesmic Desktop informs you in the bottom-righthand corner of new updates, new direct messages, and new replies. The application itself supports multiple customizable panels that let you easily read updates from different groups of users (for example, perhaps you want to categorize customers in one panel and put your family members' updates in another) and search terms (such as mentions of your name or product). Seesmic Desktop also lets you manage multiple Twitter accounts in one single application, so if you maintain a company-centric account in addition to your personal account, you can easily read and respond to comments in a single interface.

TweetDeck (http://www.tweetdeck.com)
> TweetDeck is an Adobe AIR application that takes Twitter feeds and breaks them down into several customizable "decks," or panels. Like Seesmic Desktop, TweetDeck lets you monitor a search for a term in one panel, your direct messages in another panel, and your entire Twitter stream in a third.

TwitterFox (http://www.twitterfox.net)
> TwitterFox is a way to access Twitter via Firefox, but not necessarily by navigating to Twitter's home page (*http://twitter.com*). That's because it is integrated with your Firefox browser. TwitterFox adds an icon on the status bar of Firefox that informs you of new updates, in addition to a text input field that lets you type in a new tweet.

URL Shorteners

Given that you're limited to 140 characters on Twitter and you may come across a URL that is longer than that as you traverse the Web, you'll run into a problem if you try to share the lengthy URL on Twitter. Therefore, numerous URL shortening services have been launched.

cli.gs (http://www.cli.gs)

cli.gs is a short URL service with analytics. Simply register for an account on cli.gs (so that you can access detailed analytics and statistics later) and create a Clig. Your lengthy URL will likely be converted into something like *http://cli.gs/248Af* (or a URL of your choosing with the alias feature), which is a lot shorter than the original URL. Once those links are out in the wild, log back on to the administrative console to see detailed statistics (Figure 6-7), including a chart of your total hits, a map of where the traffic is coming from, your referrer statistics, search engine crawler activity, and social media mentions.

TinyURL (http://www.tinyurl.com)

One of the more popular URL shorteners is TinyURL. TinyURL lets you instantaneously create a short URL with an alias if you desire (no login is required).

bit.ly (http://www.bit.ly)

bit.ly is another URL shortener that resembles cli.gs. Every link shared on bit.ly has its own informational page that is tied to your account (should you choose to sign up) that lets you view detailed statistics, such as the number of clicks, where they originated from, and related conversation. Moreover, bit.ly lets you see which other users on the service have shortened the same URL.

Twitter Trends

On Twitter, a lot of the same topics are discussed frequently. Finding the trends on Twitter is easy with the following myriad tools, which allow you to spot the popular stories of a given time period.

twitt(url)y (http://www.twitturly.com)

Twitturly tracks URLs in real time as they are being discussed on Twitter. Navigate to the page and you'll see exactly how many active tweets refer to a specific story. Twitturly lists the accounts and usernames of the active tweeters and also lets you share the story with your followers. Twitturly is a good tool for finding users with similar interests as you— especially as determined by them linking to a story you wrote or talked about.

Twist (http://www.twist.flaptor.com)

What do people talk about on Twitter? Twist (Figure 6-8) shows a graphical chart of what people are talking about and when. You can also view most recent tweets about the specific topic of your search.

Twitscoop (http://www.twitscoop.com)

Twitscoop shows what's hot on Twitter right now. Unlike the previously mentioned tools, Twitscoop merely takes a single phrase and highlights it. You can mouse over the phrase to see what people are saying, or if you click on the phrase, you will see a graph of user activity and view the most recent user tweets related to that phrase. The Twitscoop website is updated in real time, so as soon as you visit, you will likely see phrases shrinking in size, growing in size, reappearing, or disappearing, all depending on the current activity on Twitter.

Total Hits

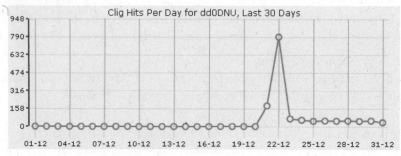

Total number of hits to this clig is **1435 hits**.

Hits World Map

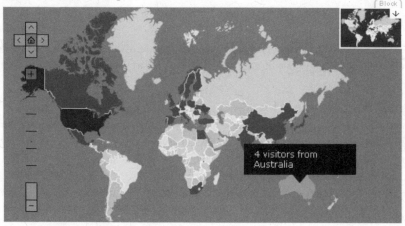

Referer Statistics

(No referer): 968 hits.
http://twitturls.com/: 33 hits.
http://twitter.com/home: 208 hits.

FIGURE 6-7. Detailed Cligs statistics

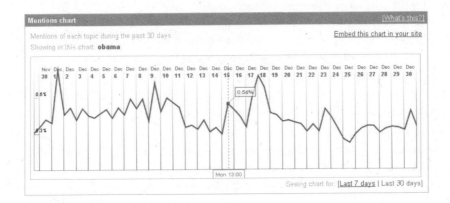

FIGURE 6-8. *This Twist chart shows the percentage of Twitter discussions specific to President Obama*

Twitter Personal Statistics

Are you interested in learning about your popularity on Twitter? Perhaps you want to learn more about someone else and his influence on the site. What about finding out how active or addicted you are to the service? There are a handful of really fun tools that will let you learn more about yourself and others on Twitter.

TweetStats (http://www.tweetstats.com)

TweetStats gives you username-specific statistics about your account. How many tweets do you post daily? How many replies do you make? What is your preferred Twitter interface (twhirl, web, SMS, etc.)? What time are you most active on the site? TweetStats reviews your entire account history and tells you all about yourself. It also creates a "Tweet Cloud" of the terms you talk about the most on Twitter.

TwitterCounter (http://www.twittercounter.com)

TwitterCounter graphs how many followers you have. By default, you will see your last seven days of followers. You can also view a 30- or 90-day graph. TwitterCounter also predicts how many followers you will have at the end of the day, based on your current rate of acquiring new followers.

Twitter Grader (http://twitter.grader.com)

Twitter Grader measures your Twitter authority and reach. The grade is calculated from the number of followers you have, the number of people you are following, the frequency of your updates, and the completeness of your profile, among other factors. In some instances, Twitter Grader suggests usernames of individuals you should follow.

twInfluence (http://www.twinfluence.com)

twInfluence shows the influence of a Twitter user based on the reach of his followers. If, for example, you had 300 followers who each had 0 followers, you are likely less powerful than someone else who has 3 followers who have 500 followers each.

Twitterholic (http://www.twitterholic.com)

Twitterholic aggregates statistics over a long period of time to tell you exactly when a user joined the service and how many followers he has accumulated over time. Twitterholic also lists the top users by number of followers and suggests additional users to follow.

TweetPsych (http://www.tweetpsych.com)

TweetPsych is an interesting Twitter application that builds a psychological profile based on the linguistic analysis of your last 1000 tweets.

TweetDeck (http://www.tweetdeck.com/beta)

TweetDeck, discussed earlier in this chapter, released an iPhone and iPod Touch application in June 2009. Its mobile counterpart offers very similar functionality to the desktop application, but also synchronizes data from your desktop to your mobile. It also offers multiple columns that can be accessed with the swipe of a finger.

Tweetie (http://www.atebits.com/tweetie-iphone)

The $2.99 Tweetie application offers a simple interface that combines all Twitter messaging (direct messages, favorites, custom searches, etc.). Tweetie also lets you look up users and add your geolocation to tweets.

Searching for People to Follow

Do you want to locate specific businesses or celebrities who use Twitter so that you can broaden your network? A number of excellent utilities work alongside Twitter to help you find these people and also figure out who else to follow.

Twellow (http://www.twellow.com)

Twellow is a "yellow pages" for Twitter professionals. Twellow lets you find professionals in just about any company. Categories include textiles, electronics, publishing, financial services, technology, and more. It is a critical tool if you are aiming to use Twitter to network with business professionals in your niche.

Twibs (http://www.twibs.com)

Twibs allows you to locate businesses that are active on Twitter. The service was launched in early 2009 and monitors thousands of businesses in every single sector. You can see how many followers a specific business entity has and follow them as well.

CelebrityTweet (http://www.celebritytweet.com)

Twitter is popular among celebrities, too. Find your favorite celebrity on Twitter using this service.

Muck Rack (http://www.muckrack.com)

With widespread adoption comes journalists looking for stories on the microblogging site. Muck Rack aggregates a list of journalists using Twitter from high-profile news sources, such as *USA Today*, the *Los Angeles Times*, FOX News, and the BBC.

Twubble (http://www.crazybob.org/twubble)

Twubble focuses on networking: the tool looks at your friends and suggests other people for you to add based on who they follow.

MrTweet (http://www.mrtweet.net)

Mr. Tweet is known as one of the most accurate tools to help you find relevant people to follow on the service. All you need to do is follow MrTweet on Twitter. It will aggregate your broadcast stream and compare it with other users and then provide you with recommendations.

WeFollow (http://www.wefollow.com)

WeFollow is a user-powered Twitter directory. Are you an expert in Buddhism and personal training? Simply tweet "Just added myself to the *http://wefollow.com* directory under: #buddhism #personaltrainer." WeFollow dynamically creates categories and shows you who has associated himself with each. You can also seek out experts in any specific area and find more people to follow in these categories.

twtvite (http://www.twtvite.com)

So far, you have learned how to connect online with people who have expertise or fame. How about connecting with regular Twitter users face to face? Before Twitter's widespread adoption, Twitter members yearned to meet with one another in person and began creating "Tweetups" (Twitter meetups). This is now easier with the aid of twtvite, which lists meetups in hundreds of cities and lets users RSVP from the interface.

Maintaining (or Scrapping) Friendships

Do you want to find out if people you follow are also following you? Perhaps they were following you one day but aren't anymore. Was it something you said? You can find out who is following you (or not) with a number of powerful tools, and you can also see exactly what you may have done or tweeted that prompted people to add you or to scrap you as a friend.

Friend or Follow (http://www.friendorfollow.com)

Friend or Follow lets you see if a person you're following is following you back. Based on this data, you can choose whether to continue or discontinue any one-way relationships. You may just want to be patient, though, too.

SocialToo (http://www.socialtoo.com)

SocialToo is an excellent service that sends you a daily digest of new followers you have gained and followers you have lost (and the tweets that have contributed to the gain and loss of followers). It is one of the most accurate follower trackers for Twitter. SocialToo also lets you monitor activity on other microblogging services, lets you automatically follow people who follow you (if you so desire), and offers a very cool survey tool called SocialSurvey that lets you tweet questions and get answers from your followers. SocialToo offers many of the smaller features free of charge, but the daily digest of follower emails can be had for a one-time fee of $20.

NutshellMail (http://www.nutshellmail.com)

NutshellMail sends you email digests of the latest tweets from you or your friends during a specified period. Most importantly, you can find out exactly when someone added you on Twitter or quit following you. Beyond Twitter, NutshellMail also lets you view updates from your friends on Facebook, MySpace, and LinkedIn, and consolidates multiple email accounts into one single digest. Surprisingly, however, Twitter's features are pretty useful, especially if you want updates for reference in the future without having to resort to Twitter Search.

Twitter Search

Finding mentions of specific text instances is easy with some enhanced Twitter search functionality. Beyond the actual *http://search.twitter.com* page, which is Twitter's official channel, you can discover trends via hashtags and easily locate that tweet you may have sent months ago.

Hashtags (http://www.hashtags.org)

Twitter is laden with messages prefaced by an @ symbol or a # symbol. As discussed previously in this chapter, the @ symbol refers to a user on Twitter. If you click on an @ link, you will be taken to that user's page. On the other hand, the # is a hashtag that refers to a trend on Twitter. If, for example, you were active on Twitter during the time of the Mumbai terrorist attacks, you may have seen several thousand tweets containing #mumbai. On the other hand, if you were active on Twitter during the Iran elections, you probably stumbled upon hundreds of mentions of #IranElection. Terms are aggregated into hashtag trends that are visible on sites like Hashtags.org. Clicking on a hashtag on one of these sites shows related tweets and a graph of the particular trend. Hashtags are heavily used at conferences and meet-ups or to refer to newsworthy events.

TweetScan (http://www.tweetscan.com)

TweetScan is a search tool like Twitter Search that monitors users, activity, hashtags, keywords and phrases, and more. Twitter Search and TweetScan should be used in conjunction for the most comprehensive search, as each one alone may not cover all bases.

twazzup (http://www.twazzup.com)

A powerful search tool that adds real-time updates, twazzup also allows you to follow users with the click of a mouse. Hover your mouse over a Twitter user's name to view statistics about that user (bio, number of followers, and location). twazzup also shows a panel of "Featured Tweets," top trendmakers (those speaking in-depth about a particular search term), related photos, and popular links that individuals are sharing on Twitter.

Mobile Applications

When you've become addicted to Twitter, you'll certainly want to take it with you. Fortunately, there are applications that make Twitter portable, be it on a BlackBerry or iPhone.

TwitterBerry (http://www.orangatame.com/products/twitterberry)
> BlackBerry users can enjoy Twitter on their devices with this custom solution for all BlackBerry mobile devices.

Twinkle (http://www.tapulous.com/twinkle)
> Twinkle is a full-fledged social application for the iPod Touch or iPhone that lets you communicate with friends, upload photos, and even find other Twitter users with Twinkle's geolocation features.

Twitterfon (http://www.twitterfon.net)
> Twitterfon is a simple Twitter application for the iPod Touch or iPhone. It features a search tool and lets you reply, retweet, and see conversations in a separate view, among other basic functions.

Summary

Twitter's large community and its succinctness make it a viable choice for user interaction and engagement. On top of that, Twitter has helped facilitate successful business relationships and then some. Because Twitter is fast and lets you share your thoughts quickly, it has become one of the premier tools for crisis communications. For business purposes, Twitter can help raise revenue, as evidenced by Dell's success using Twitter as a medium for selling refurbished computers, and by Namecheap's success with a contest that brought an increased number of domain name registrations. Twitter is also a powerful customer service tool; both Comcast and JetBlue have successfully leveraged the system to produce happy customers, with turnaround times that are faster than the typical support phone call.

Twitter has also helped strengthen brands. Popular shoe retailer Zappos is almost ubiquitous on Twitter and has seen a positive return on its investment in the service. In fact, it's incredibly reassuring to have its CEO, Tony Hsieh, immediately accessible via the service.

Finally, Twitter can act as a powerful broadcast medium, especially for government entities. The Israeli Consulate was the first of which to hold a "citizen's press conference" using Twitter and found it a great way to convey country sentiment to so many people so quickly. The Israeli Consulate was able to receive questions immediately and provide feedback almost instantly.

Twitter has other business uses as well, such as facilitating the creation of a personal brand, getting direct feedback, and networking with individuals. Additionally, there are a plethora of tools that help you attain these business goals. Further, additional tools are available for Twitter that can help you achieve even more business objectives. This book touches upon some of these tools. Additional tools for Twitter and business are listed at *http://www.briansolis.com/2008/10/twitter-tools-for-community-and.html*.

Endnotes

1. *http://www.seomoz.org/blog/customer-service-and-reputation-management-the-twitter-way-a-case-study*

2. *http://consumerist.com/5093978/jetblue-twitter-faster-than-customer-service-rep*

3. *http://twitter.zappos.com/employees*

4. *http://www.blogs.zappos.com/blogs/ceo-and-coo-blog*

5. *http://www.pistachioconsulting.com/twitter-competition-verizon-comcast*

6. *http://www.twitter.com/gkerkvli/statuses/1055609599*

7. *http://www.lateralaction.com/articles/how-do-you-define-success*

CHAPTER SEVEN

Getting Social: Facebook, MySpace, LinkedIn, and Other Social Networks

Social networks such as Facebook, MySpace, and LinkedIn are powerful for message broadcasting and for brand awareness. Because these sites inherently connect people with similar backgrounds and interests, hub pages dedicated to products and services are often invoked either by fans or by marketers/businesses out a desire to create strong associations between individuals and products. On Facebook, the two predominant marketing tools are the application or Facebook Page; on MySpace, it is the customized profile. In this chapter, we will look at how to use these sites and how they benefit marketers aiming to expand the reach of their products.

Introduction to Social Networking Sites

In this book, you'll see several different types of social media marketing channels, from social networks to social bookmarking sites and social news sites. "Social networks" and "social networking sites" are generic terms for sites that are used to connect users with similar backgrounds and interests. Social networks are profile-based sites that encourage users with relatively comparable backgrounds to meet and initiate relationships with one another.

Social networks are some of the most popular sites on the Internet. For example, Facebook and MySpace are in the top 10, according to website statistic service Alexa. For some users, these social sites are an addiction. Employees have been criticized for spending too much time on the more popular social sites.[1] This goes to show that while social networks may have initially appealed mainly to our youth, they are now becoming a key part of the livelihood of individuals in all demographics.

As well as connecting users who have shared interests, family backgrounds, or political views, social networking sites also foster relationships based on sexual orientation, religious beliefs, or racial identities. People also connect by virtue of having similar hobbies (common favorite TV shows, musicians, and more).

The three networks we will focus on in this chapter are Facebook, MySpace, and LinkedIn. All three services are free to join, but some additional features (such as Facebook gifts and extended LinkedIn features) are available for a nominal (or sometimes not so nominal) cost. In all, the idea is to share information about yourself and connect with old friends and colleagues. In many instances, new friendships crop up as well.

The premise behind social networks is to create a profile displaying your identity that you can then share with your friends. These profiles are typically interactive and users can often comment on or share them with other friends. The profiles are fully customizable, and as the owner of your profile, you have complete control over what is displayed on the profile page. In some social networks, you can also create network pages. If you are the manager or administrator of these pages, you can make changes or amendments to them and you have full control over what is displayed.

When you join a social network, you begin by creating and customizing your profile. This need not be more than just your name and location, but depending on the site, it can include your birthday, marital status, work information, sexual orientation, religious affiliation, lifestyle habits, and favorite entertainment media. Further personalization is possible by uploading photographs and videos.

Once your profile is set up, you can connect to friends and colleagues by searching for them by name. You can also perform searches based on interests, location, schools, or affiliations. When you find an individual whose profile matches your search, you may wish to immediately initiate a friendship with her (often, you can add a friend to your network by choosing the "Add Friend" or similar option), or you can email her directly.

For those in Generation X and earlier, social networks may be perceived as a big violation of privacy. However, volunteering personal information is extremely prevalent among the youth of today. As such, social networks are often goldmines of information, but user profiles are also very detailed and exhaustive. In general, if you are concerned about the information being conveyed via your profile, there are numerous privacy settings available. Facebook, for example, has both site-wide and individual privacy settings that can be granularly targeted to groups of people or to specific users on the site. In other words, you can prohibit groups of friends from viewing specific raunchy photos, for instance, and at the same time you can show your employment information only to your closest family members.

Facebook: The Digital You

In 2008, Facebook surpassed MySpace as the most popular social network of its time. Today, the fastest-growing demographic on Facebook is the over-35 crowd, and over 200 million users are active on the service.[2] Facebook is a simply laid-out website that allows users to share information about themselves, from their names to their birthdays to their favorite television shows, and then some. Facebook also has a very heavily used photo tool that encourages "tagging." Tagging is an image-sharing tool that allows you to identify other Facebook users in your photos. Once you tag an individual, she is alerted to the existence of the photo and the photograph shows up in her personal profile. An additional feature of Facebook is its virtual gifts feature, which allows individuals to give gifts to their friends, and most gifts are valued at $1, which can be paid by credit card. Further, Facebook supports third-party applications and boasts thousands of applications with millions of users. For marketing and identity association, individuals on Facebook can also align themselves with many different companies and brands by interacting with existing Facebook fan pages or by creating one from scratch.

Due to the variety of features available, Facebook seems poised to grow in the future. Forever innovating with multiple destination pages (gifts and photos are popular choices), Facebook is a common term in the American household and in countries abroad.

There are different ways to use Facebook as an advertising medium. It's important to know, however, that the initial community of Facebook users came to the site with the intention of building upon existing personal connections. Only later did marketers engage with the Facebook community, when they realized that there is a whole lot of advertising potential on the site. Therefore, it's important to tread carefully. If you are initiating contact with your business connections for matters that are strictly business, it may be received better than if you initiate a friendship with an old high school friend and immediately try selling your real estate service to him. Social media etiquette using Facebook is more important than ever. (See Appendix A for some guidelines for social media etiquette.)

Profile Customization and Blog Posts

Your can personalize your Facebook profile with your group affiliations, hobbies, and applications you choose to install. You can also port your own blog posts to Facebook via its internal syndication tools. To do this, click Links on the left side of your home screen, then post a link and add a comment. The link and comment will be displayed on your profile page and in the "What's On Your Mind?" area (see Figure 7-1).

When your friends log in, they will see your posted items in the News Feed section on their home screens. The Facebook news feed captures friends' recent activity and status updates, such as, "John Smith is headed to bed," or, "Amy Clark will be participating in the NYC Marathon," with a link for more information. These are interactive updates that Facebook friends can "Like" (essentially indicating that they like the update) or comment upon. The

FIGURE 7-1. *Facebook's posted items*

news feed effectively broadcasts your updates to your friends' home pages. Therefore, if you update your profile frequently, you'll always be on your friends' radars.

When you log in, you will also see Facebook Highlights on the right-hand sidebar. Highlights show events that your friends are attending, photos that your friends have uploaded (that may have been Liked by other friends), and product pages that your friends may have become fans of. Facebook Highlights effectively show you how users are interacting with other Facebook tools and features, and give you insights into what your friends are doing on Facebook beyond their status updates.

Other Facebook features on the righthand sidebar include friend suggestions (Facebook determines who you may know based on who your friends may know), fan suggestions (have six of your friends just become fans of Dave Matthews Band? Maybe you like that band also), and upcoming birthdays and events.

Additionally, since Facebook supports third-party tools, many companies have jumped at the opportunity to use the site as a one-stop shop for sharing content. Twitter, for example, has a popular Facebook account with nearly 100,000 regular users (*http://www.apps.facebook.com/ twitter*). The NetworkedBlogs application (*http://www.facebook.com/apps/application.php ?id=9953271133*) automatically ports your blog posts to your profile and allows bloggers to network one another. Enabling these tools on your Facebook account is as simple as locating them and selecting Allow. You must authorize the access requested in order to get full benefits of the features (Figure 7-2).

FIGURE 7-2. Allowing access to third-party applications on Facebook

The Facebook "Product" Page: Aligning with Your Favorite Brands

By far, the most popular free tool for marketers on Facebook is the Facebook Page (*http://www .facebook.com/pages/create.php*). Facebook Pages are profiles for business, products, services, your pets, and even public figures. They allow entities to have a real presence on the social network and are designed just like personal profiles. There are 62 different categories for Facebook Pages, from convention centers/sports complexes to local grocery stores to rental cars and websites on the brand/product level to models and actors on the public figure level. If you have something to market, chances are that Facebook has already thought of a relevant category and will allow you to create a page in order to publicize it on Facebook.

Personalizing your Facebook Page

Once you've selected a category for your Facebook Page, it's time to add content. You can customize this page with a profile picture and information that is relevant to your visitors (for example, birth date or date the company was founded, plus a biography or a company overview). Facebook Pages are also equipped with Facebook's basic applications, which require no extra third-party access or authorization. These applications include Discussion Boards, Information, Notes, Photos, Posted Items, Events, and the Facebook Wall, which lets people display and view public messages on your page (which you can later delete if necessary). If you want, you can add third-party applications to a Facebook Page, including Facebook's recommended RSS Connect and YouTube Video Box applications for business, among thousands of other applications that can add more personality and value to your page.

Sharing your Facebook Page with your company

After creating your Facebook Page, it's time to publish it and make it public. This is a quick and painless process and is the last step you'll take before you actively start telling people it. Ideally, you'll want to market it internally first and get your company on board. You can add administrators to the profile; these individuals can then market under the Facebook Page profile name. For example, if you are creating a page for a company named ACMEWeb, any administrator who interacts on the profile page will be doing so as "ACMEWeb" and not as "George Burnett" or "Melinda Foster." This way, you are clearly representing your company on the profile page and there's no confusion (though internally, you may want to delegate who handles these communications, as it may be unclear where the messages originated from when there are multiple administrators). Anyone with administrative access can see detailed statistics and add and remove applications.

Sharing your Facebook Page with the community

After your internal employees have joined, you should broadcast your Facebook Page's existence to the world. You should leverage existing channels for this request: your company newsletter, your Facebook contacts (but only those you know would be interested), email messages to business colleagues, or a blog post on your site. You can also offer incentives for users to join, as Sears does by giving a $10 coupon to anyone who becomes an official fan of the Sears Facebook Page and clicks the coupon link (Figure 7-3).

If you click on the "Become a Fan Today! Click Here to Get Your Free $10 Sears Coupon Now" image shown in Figure 7-3, you get redirected to sears.com, where you confirm your Facebook information to be eligible for your coupons (Figure 7-4). With this implementation, the company can verify if you are truly a Facebook Fan and reward you accordingly.

It's also interesting to note that Sears has customized its Facebook page pretty heavily with personalized applications; Facebook Pages can have a lot more than the basic boring profile that your friends have used.

The best-performing Facebook Pages are those that provide compelling and engaging content. For companies, it is a good idea to provide utility for your users; engaging them with relevant content is ideal. Sears did this by linking to a coupon and also by promoting some other company-specific charity marketing initiatives. In general, you can consider linking to your

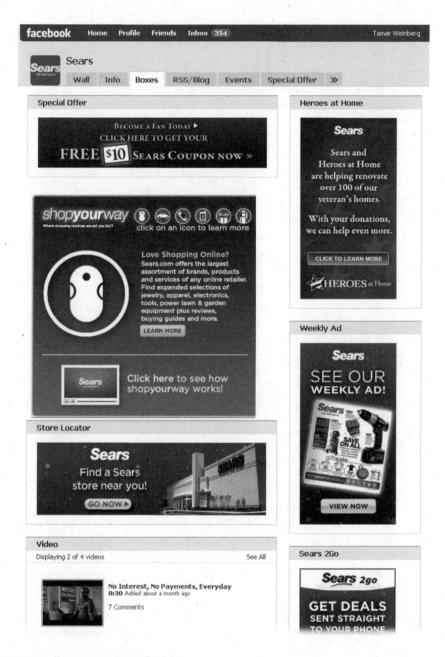

FIGURE 7-3. Sears offers incentives for Facebook fans

company blog, serving articles or whitepapers relevant to your audience, or by showing related sales and promotions.

FIGURE 7-4. Sears Facebook promotion for $10 in coupons

Just as profile pages can be updated at any time, Facebook Pages can also be updated with new photos, new wall posts, new updates, and new applications whenever necessary. The downside of Facebook Pages is that users will not be notified if there are new postings on the page, such as new events or moderator announcements via the discussion boards. In fact, even official updates to the Facebook Page are not sent to the user's inbox. Official updates can only be viewed by clicking on the "View Updates" link on the top-righthand corner of the page. At this time, the only Facebook Page communication allowable on the service is an email that you can send to your contacts suggesting that they become a fan of your Facebook Page.

However, Facebook Pages can offer detailed statistics—all free of charge—about who is interacting with your profile. This tool, called Facebook Insights (Figure 7-5), shows the number of page visits per day, demographic information, and popular pieces of content.

Armed with this information, you may want to go the paid route and target your readers with ads. These ads are useful because Facebook users typically provide very detailed and up-to-date demographic information, so Facebook can provide a powerful platform for advertising.

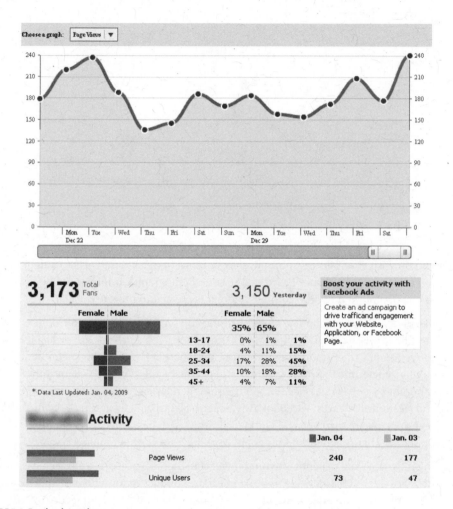

FIGURE 7-5. Facebook Insights

Further, the Facebook Lexicon (*www.facebook.com/lexicon*) highlights trends that are obvious across Facebook Walls, equipping advertisers with even more data that can assist with an advertising campaign.

Paid Advertising on Facebook

Although there are plenty of free marketing options available on Facebook, paid advertising is a viable marketing tool, so we will touch on it briefly here.

The Facebook advertising page (*http://www.facebook.com/advertising*) outlines the process for ad creation (Figure 7-6). Choose the URL that you want to advertise, create the ad and add

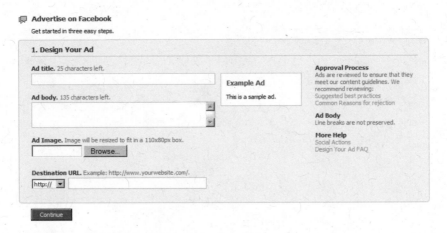

FIGURE 7-6. Creating a Facebook advertisement

a photo if you wish (nonphotographic ads may perform better, though, as they are less intrusive), choose your audience based on the site's demographic information, price your ad (cost-per-click), and then review your ad before it is pushed to the site's users.

Facebook Ads are the most popular types of ads on the service and are visible on the sidebar of most profiles and Facebook Groups. More comprehensive Social Ads are available and are best for larger brands to promote their products. With both advertising opportunities, you can target specific geographic locations (country and even city, all of which is determined by IP address and not by network association on the site), gender, age, education status, political views, and relationship status. Facebook Ads can also be targeted to specific keywords that are visible on profile pages and even the workplace, which can be particularly helpful if you're offering a service that is beneficial for people working for a certain employer.

Ad pricing is variable and depends on your budget constraints. Facebook's ad pricing is flexible, with costs starting at $0.01 per click (with a minimum budget requirement of approximately $5 per day). It is recommended that you begin your advertising with a budget of $250 at first, but depending on performance, you may want to make adjustments to this amount.

To assess the performance of your ad, you can use Facebook Insights, which can provide you with detailed metrics, such as the number of impressions delivered, the total number of clicks, the click-through rate (CTR), the average cost per click (CPC), and more.

You can find detailed information regarding the variety of Facebook Ads and the metrics at *http://www.subliminalpixels.com/2008/12/06/how-to-set-up-and-track-facebook-ads*.

Facebook Groups, User Profiles, or Fan Pages?

Facebook Groups provide yet another way for individuals to align with a cause, product, business, or brand, though there are some limitations to group engagement. Very similar in presentation to Facebook Pages, Groups are much easier to set up and customize. The major limitation of Facebook Groups is that they are not as customizable as Facebook Pages. For example, there is no support for third-party applications in Facebook Groups. With Facebook Groups, interactions are only possible on the Wall, via discussion boards, photos, posted items, and videos. Facebook Groups uses an older technology; the Facebook Pages feature was introduced several years later and consequently supports more features.

The bottom line is that Facebook Groups are limited in scope. You cannot add applications to a Group. You cannot view detailed statistics and metrics about the users who interact with a Group.

Some companies are not aware of the existence of Facebook Pages and instead create personal accounts under the company name. This strategy violates Facebook's rules, though many individuals and companies have been able to successfully create these types of accounts, since the Facebook staff does not police these infractions very rigorously. Still, it's frowned upon by users (after all, who wants to be friends with a guy whose full name is supposedly "ACMEWeb Inc?"). This strategy is not effective for a variety of reasons: if you are looking for increased branding, Facebook Pages provide the most ideal solution because they are user-searchable and publicly accessible by search engines. On the other hand, user profile pages are private by default. Additionally, Facebook personal profiles are limited to 5,000 friends, so if your brand grows (or if your celebrity status blossoms), you will be unable to confirm additional incoming friendships and they will be unable to add you as a friend. Fortunately, with a Facebook Page, you do not need to consistently approve fan requests the way you have to approve friend requests on a personal profile; people can join your Facebook Page without your manual approval. Facebook Pages appear to be the smart way to advertise products and services on Facebook. While it has some shortcomings (inability to email users directly), the Facebook Pages feature has more robust functionality and is much more engaging for Facebook's user base than Facebook Groups or profile pages.

Facebook Applications for Marketing

Another way to market your brand or business is by engaging the community through the creation of Facebook Applications. These are third-party applications that any developer can port to Facebook with a basic understanding of the Facebook coding language (Facebook Markup Language, or FBML). These applications are visible in the "Boxes" portion of a Facebook Profile and can add a little more personality to a user profile or Facebook Page.

BURGER KING'S ATTEMPT AT BRANDING THROUGH FACEBOOK

In the beginning of 2009, Burger King released a clever Facebook application called the "Whopper Sacrifice," through which Facebook members were told that they could get a free Whopper at their local Burger King when they removed 10 friends from their Facebook accounts.

The application gained momentum and nearly 234,000 friends were sacrificed in thousands of attempts to score a free meal. After all, who could resist the opportunity to capitalize on a poor economy?

Within 10 days, however, Facebook requested that Burger King pull the application because Whopper Sacrifice violated Facebook's protocol. The idea behind Facebook is to keep and maintain your friends, not to sacrifice them.

However, in the end, the Whopper Sacrifice was a great success for Burger King. Over 13,000 blogs referenced the campaign. There are over 142,000 results about Burger King's branding attempt on Google's search.

Initially, Facebook Applications were very well received. Today, they are not as heavily used due to the sheer number of available applications that fundamentally perform the same tasks (take a lesson from Burger King's Whopper Sacrifice application and do something different!) Further, Facebook encourages users to share applications with other users who have not yet installed them, but many people are hesitant to respond to the numerous requests they receive to install new and unknown applications.

When executed properly, Facebook Applications may benefit companies looking to spread a marketing message. It's advisable to create an application that is useful and addresses a specific need. For example, the NetworkedBlogs and Twitter applications mentioned earlier in this chapter have allowed users to connect other services with their Facebook profiles. If you are selling multiple products and have a storefront, building an application that allows users to create custom wish lists may be a viable solution; on the other hand, if you are hosting a sports site, you can capitalize on sporting events (for example, March Madness) to encourage users to choose their winning teams by offering the highest-scoring individual a prize.

Success with Facebook Applications comes with doing something different, but also from providing added value to the user. Freebies and incentives never hurt, either.

MySpace: Personalized Connections

MySpace is a popular alternative to Facebook. Considered one of the first public social networks, the site gained traction in 2004 and was used heavily to encourage musicians to broadcast their works to new users. MySpace's profile pages (Figure 7-7) are far more customizable than Facebook's: users can choose their own background designs. Further, MySpace makes it easy to share music with the public; you can upload videos to the My Music portion of your profile, and when other users access your profile, the music plays automatically. MySpace, in essence, allows you to add personality to your profile page.

FIGURE 7-7. A typical MySpace profile page

When users sign up on the site, they volunteer the same kinds of personal information that they would on Facebook. Users can write brief biographies and share their favorite TV shows, books, and movies. MySpace asks users to share other personal details, such as their social habits (smokers or drinkers), whether they want or have children, their occupations (in general categories), and household incomes. On MySpace, users can also add the schools they have attended, as well as former and current locations and employment roles. MySpace users can also affiliate themselves with specific networking categories (for example, Music Marketing or Dance Production).

For the most part, MySpace works very well as a social network that lets you meet new users. It does not support third-party applications nor does it have company-specific pages. Instead, the most effective forms of marketing are via profile personalization with custom backgrounds and effective coding.

Without applications and other destination pages, MySpace marketing mostly takes the form of profile promotion. A user can customize his profile as much as he wishes; there are thousands of MySpace profiles and themes available, but an able web developer with knowledge of CSS can create a profile that will be unique, fun, and engaging.

Creating Your MySpace Profile

The MySpace profile is where you make your brand shine. Use a memorable picture as your profile photo. Add character to your profile by choosing a background image that makes sense for your brand; avoid busy imagery that will detract from your mission. Add a song or music soundtrack only if it is highly relevant to your brand; otherwise, skip it. You can also limit the types of comments users can write on your profile by disabling HTML and other coding. This will come into play as you get more and more active friends on the site.

Adding a Blog to Your Profile

MySpace also allows users to create their own blog entries. The blog entry is simple and it comes equipped with a familiar WYSIWYG editor. One of the most interesting features of the blog entry feature is that it lets you add links to podcasts. This lets musicians and podcasters broadcast the fact that they have added a new blog post and also gives users the opportunity to listen to the podcast.

Networking and Broadcasting Your Profile to the World

Once you have created your MySpace profile, you will want users to know about it. You can do this by networking with the right people; extensive searches may be useful and necessary. Some tools (not endorsed by MySpace) allow any MySpace user to find and add hundreds of users at a time; this, in turn, raises awareness about his profile as more and more people confirm the friendship. Many marketers will encourage the use of these applications, but use them sparingly to avoid being banned by the service.

You can also expand your network by hand. MySpace's search feature allows you to search by gender, age, relationship status, location, ethnicity, body type, height, and lifestyle (religion, social habits, sexual orientation, and education background, among others). While this information can be helpful, many MySpace users do not provide accurate profile information (for example, young children on the site may choose to represent themselves as much older out of safety concerns).

Don't Be Forgotten!

Once you get friends on MySpace, make sure you're remembered. Send out bulletins to your network of friends once in a while (don't be excessive) and be sure that these bulletin messages are relevant to your friends' interests but also related to your mission on the site. You can cross-post bulletins to your blog, which will help newcomers discover your involvement and perhaps prompt them to initiate a friendship with you. Sending out bulletins is also a good way to accumulate page views on your profile page.

Video marketing is also beneficial on MySpace, so be sure to occasionally share videos that are related to your niche and provide utility to your connections. You can only post one video on your profile page at a time, so make sure you update your network (via blog post or bulletin) when you swap out an old video for a new one. It may be ideal at this point to upload your videos to another site, such as YouTube or MySpace Video (*http://vids.myspace.com*), for additional marketing exposure. We'll discuss video marketing in more depth in Chapter 11.

To keep your friends intrigued, be communicative, even if the conversation is not relevant to your business. As with any other social network, being genuine and open, even if it doesn't fulfill your marketing objectives, is key.

Is MySpace Marketing Worth It?

Of the social networks, MySpace is probably the least capable of marketing fulfillment. Profiles can be created anew, but the site has not been evolving as heavily as Facebook has in the past few years. Facebook's Pages and Applications are relatively new and popular additions to the service, whereas MySpace lacks any new attractive functionality beyond the profile page. This is also why Facebook has gained the edge in the online world; new features and heavy user engagement have made Facebook a more viable long-term choice as a social network with potential personal and marketing benefits.

Despite the shortcomings of MySpace for general marketing, it is probably one of the best tools for music marketing, especially for bands looking to branch out and be recognized. Event promoters often use MySpace to find local talent to perform at nearby venues, so musicians and bands are encouraged to use the site to advertise their gifts to a broader audience.

Getting Professional with LinkedIn

As a professional niche social networking site, LinkedIn is best used to connect with professional colleagues (former and current) and with individuals in the same or related industries to establish relationships and to find recommended services. LinkedIn boasts over 40 million experienced professionals around the world in over 150 countries.[3]

LinkedIn is best for finding potential clients, service providers, subject experts, and colleagues. The service can connect professionals for business opportunities and jobs as well, and many recruiters and other job seekers use the site to find new careers.

LinkedIn works similarly to Facebook and MySpace in terms of registration. New users sign up and complete their profiles like a resume. LinkedIn encourages profile completeness: it urges users to complete their profiles to 100% by filling out educational background and past jobs, and by getting recommendations from colleagues who are in the same user's network.

Once your LinkedIn profile is somewhat complete, you should start connecting with individuals to build your network connections. Feel free to add family, colleagues, coworkers, and old school friends. On LinkedIn, you may need to have your colleagues' email addresses to establish them as contacts, especially if they are not in your immediate network. If this information is not readily available, you can usually find another means of connecting with them, whether by groups and associations (to be discussed later in this chapter) or by past businesses (though you will need to specify a job description where the relationship occurred). As a precaution, you should add only people you know; if too many people click "I don't know Frank Paulson" too many times, you can be penalized on the service.[4]

The most successful business users of LinkedIn focus on providing professional services. While LinkedIn is a viable business tool, it is not the most ideal instrument for mass marketing. However, there are a variety of methods that can help bring LinkedIn closer to the consumer, as long as that consumer is a registered professional on the service.

LinkedIn Answers: Showing Off Your Expertise

One of the most effective channels within LinkedIn is the Questions and Answers section (*http://www.linkedin.com/answers*), which connects professionals and provides them with marketing strategies and tactics for improving their outreach efforts (Figure 7-8). Benefits of utilizing LinkedIn Q&A include relationship building, idea generation, brainstorming, and establishing thought leadership. To make the most of LinkedIn Answers, the first step is to build your network. The more connections you add, the stronger your network will become. Your overall LinkedIn network is three degrees deep, so you will be exposed to questions from direct connections, second-level connections (friends of your direct contacts), and third-level connections (friends of your second-level contacts). Therefore, your LinkedIn network can grow exponentially, depending on how many connections you build and how many connections they have.

Your LinkedIn Answers network is also three degrees deep. Depending on your engagement on the site, you may be able to utilize these connections for future outreach, growth, and development.

Asking questions is simple. You type your question, add details, and categorize it. You can also ask questions of selected contacts in your network. Questions can be geotargeted to country and postal code. Finally, you can specify whether your question is related to recruiting, promoting your business/service, or job seeking.

If you're asking questions of your network, it's also helpful to start answering them in your network. The more questions you answer, the more opportunity you have to establish yourself as a LinkedIn expert. Hundreds of users are actively engaged on LinkedIn and use LinkedIn Answers heavily. The most popular expert in my network has answered 177 questions. Additionally, individuals who ask questions are able to elect the best answer for any particular

FIGURE 7-8. LinkedIn Answers lets you show off your expertise

inquiry, so individuals can establish credibility by giving the most engaging and valuable answer.

Fortunately, LinkedIn's Answers database is very robust and has a lot of great questions already archived for your perusal. You can get ideas and inspiration from existing (and answered) questions by searching for specific terms. Additionally, you are encouraged to participate in the discussions that relate to your business and industry.

For customer acquisition, it is possible to leverage your professional retail contacts by asking the right questions. You can get some interesting insights when your contacts and their networked connections engage in an ongoing dialogue. Plus, it's a great way to raise brand awareness, especially if you identify yourself as a representative of your company. The question and answer shown in Figure 7-9 shows this phenomenon and how brand awareness is actually possible via LinkedIn Answers.

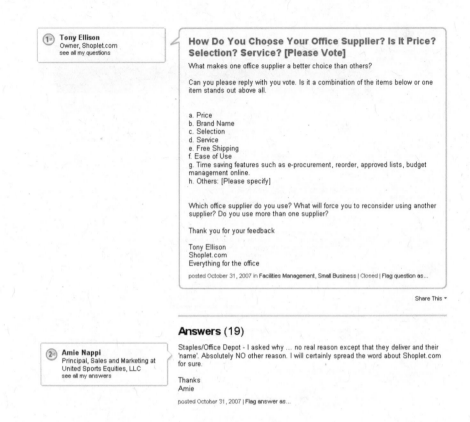

Tony Ellison
1st
Owner, Shoplet.com
see all my questions

How Do You Choose Your Office Supplier? Is It Price? Selection? Service? [Please Vote]

What makes one office supplier a better choice than others?

Can you please reply with you vote. Is it a combination of the items below or one item stands out above all.

a. Price
b. Brand Name
c. Selection
d. Service
e. Free Shipping
f. Ease of Use
g. Time saving features such as e-procurement, reorder, approved lists, budget management online.
h. Others: [Please specify]

Which office supplier do you use? What will force you to reconsider using another supplier? Do you use more than one supplier?

Thank you for your feedback

Tony Ellison
Shoplet.com
Everything for the office

posted October 31, 2007 in Facilities Management, Small Business | Closed | Flag question as...

Share This ▾

Answers (19)

Amie Nappi
2nd
Principal, Sales and Marketing at
United Sports Equities, LLC
see all my answers

Staples/Office Depot - I asked why ... no real reason except that they deliver and their 'name'. Absolutely NO other reason. I will certainly spread the word about Shoplet.com for sure.

Thanks
Amie

posted October 31, 2007 | Flag answer as...

FIGURE 7-9. Asking targeted business questions on LinkedIn

Recommend Experts

LinkedIn also encourages professionals to recommend service providers for a variety of business-related tasks. The LinkedIn Service Provider Recommendations Engine (*http://www.linkedin.com/svpRecommendations*) allows you to find experts in a variety of disciplines (Figure 7-10). The higher the number of recommendations, the more visibility the service provider receives on LinkedIn. Note, however, that LinkedIn emphasizes recommendations of individuals and not retail outlets or business entities. Therefore, this is a great tool for the one-man business or artist, but it may not work for George Burnett as a representative of ACMEWeb, especially if George Burnett moves on to another position at another company.

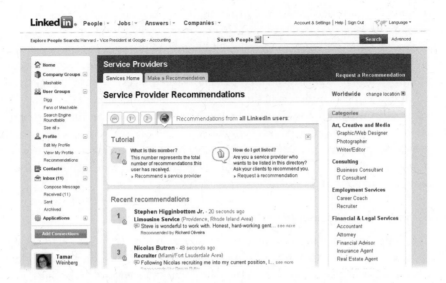

FIGURE 7-10. LinkedIn's Service Provider Recommendations Engine helps you locate recommended experts

LinkedIn Service Provider Recommendations is a great but not heavily trafficked tool that can help you locate providers who can assist you with your business needs.

LinkedIn User Groups: Connecting Users Based on Interests and Associations

One of the biggest and most popular features for LinkedIn community building is the LinkedIn Groups feature, which connects numerous individuals with similar interests. Popular groups on LinkedIn are centralized around communities or organizations (for example, Friends of UNICEF) and events (for example, Boston Marathon Finisher), though there are very few LinkedIn groups associated with brands that exist to connect consumers with the businesses they are associated with or passionate about. For example, the visible Amazon and IBM brand groups are dedicated to former and current employees (it is possible for nonemployees to join, but moderators may not necessarily approve the membership request).

While the current existing groups on LinkedIn are not heavily used for brand awareness, a number of sites (mostly in the social web sphere) have engaged their users via LinkedIn and exist simply to connect users with an interest in those products. For example, the Digg LinkedIn group boasts over 300 members; the Mashable.com social news website has over 1,800 members. It may be the case that many businesses do not yet realize the potential of using LinkedIn Groups for brand awareness.

Considerations when creating a LinkedIn group

Groups are easy to set up and require nothing but a group name, a small and large logo, a group summary and detailed description, a group type designation (alumni group, corporate group, conference group, networking group, nonprofit group, professional group, or other type of group), and a website. An owner willing to handle management of the group should provide his email address. Group visibility can also be determined on this page (e.g., whether the page is publicly accessible in LinkedIn's group directory and/or whether to allow group members to display the group logo prominently on their LinkedIn profile pages). Groups, like answers, can be geotargeted to a single location, and they can also be targeted to a specific language. If your group is sure to grow quickly and if you do not wish to manually approve members, it may be beneficial to be lax on the Group Access permissions and allow users to join without approval from a group owner. Even for small groups, this may be ideal, as users who are impatient to join your group may start sending requests directly to your inbox. Management of these requests can be cumbersome.

Communication within LinkedIn groups

Once your group is set up, you can wait for users to start joining or you can start telling your colleagues and friends about it. On LinkedIn Groups, there are discussion boards and user-contributed news articles. Group owners can also import RSS feeds from relevant sources directly to the group's home page. Email messages are sent to group subscribers (by default, until they opt out), so more conversations and discussions contributed to LinkedIn groups translate to more potential for targeted email messages, and consequently, brand awareness and engagement.

As a group owner, you can target group members by viewing their email addresses and exporting them to a file. This is another way to tap into your social media audience. Note that this facility should not be abused; if you are targeting your ACMEWeb brand profile on LinkedIn and choose to email your contacts offsite, be sure to identify yourself as the group owner of the ACMEWeb LinkedIn group. Unless the relationship is explicit, recipients of the email may feel that their freedoms have been violated.

Group owners can also connect directly with members in the same group and see their interests and profiles, as long as they are within the first three degrees of connectivity in your LinkedIn Network.

LinkedIn Company Groups

Internal communication is possible with LinkedIn Company Groups, an extension of the LinkedIn Groups functionality for users, with the added benefit of connecting company employees. LinkedIn Company Groups are not used for marketing, but they are explained here to prevent confusion between Company Groups and User Groups.

LinkedIn Company Groups will only permit current employees of a particular company to be members of the group. LinkedIn is very explicit about this, and allows you to easily remove anyone who is not a current employee from the group (and to report the user to LinkedIn in the event of abuse). Like User Groups, LinkedIn Company Groups support News Articles, but instead of a discussion board, LinkedIn Company Groups have a Q&A tab to encourage employees to discuss internal affairs among themselves. Collaboration and communication are encouraged as well.

LinkedIn Applications

In late 2008, LinkedIn announced the debut of LinkedIn Applications to allow professionals to share information about themselves and to collaborate with others. Compared to Facebook, only a small subset of applications is available to the public, including Amazon reading lists, WordPress blogs, polls, Box.net (for file sharing), Huddle workspaces, blog links, company buzz mentions, a travel itinerary, and two presentation platforms (Google and SlideShare). While companies cannot create third-party applications to start a widespread marketing campaign (as on Facebook), these applications can assist individuals looking to market themselves via their profiles. These applications can be easily embedded onto a user profile, and those interacting with the profile can get more information about the individual. In this way, LinkedIn Applications can also add personality to a user profile.

LinkedIn DirectAds

The LinkedIn DirectAds feature provides another marketing opportunity (*https://www .linkedin.com/directads*) in the form of paid advertising. Advertisements are contextual and can be targeted to certain types of professionals (as determined by company size, job function, industry, seniority, gender, age, and geography). DirectAds come at a fixed cost of $10 cost per mille (CPM, $10 per 1,000 impressions).

Like Facebook Ads, DirectAds are not very intrusive. Users can specify a short headline followed by a two-line description. LinkedIn allows the display URL to differ from the actual URL, which is particularly useful if you want to link users to a lengthy URL, but only want the user to see a short, memorable one.

After you craft your ad (otherwise known as the "creative") and start running your ads, you can view statistics about your ad performance on the Ads Dashboard and Ad Details pages.

The Big Social Networks Abroad

Facebook, MySpace, and LinkedIn are the popular social networks in the United States, but there are hundreds of popular networks in countries abroad. In this book, the most popular of these networks will be introduced, but not expounded upon.

The following sites are the primary players in the social networking sphere:

Friendster (http://www.friendster.com)
> Friendster was the first major social network in the United States. Founded in 2002, before MySpace and Facebook were even visions in the minds of their creators, Friendster initially gained much momentum. However, with Friendster languishing in development updates, MySpace became the popular social network and surpassed the company in 2004. Friendster gained popularity in Asia, particularly in the Philippines, and currently boasts over 90 million members.

Orkut (http://www.orkut.com)
> Orkut is a social network owned by Google and is popular in Brazil (and to a lesser extent in India). Today, Orkut boasts over 40 million members, the majority of whom are between the ages of 18 and 25.

Bebo (http://www.bebo.com)
> Bebo is a social networking site that is popular in the United Kingdom, Australia, and New Zealand. Bebo has many extensions, including Bebo Music, which allows artists to share their works digitally, and Bebo Authors, which allows writers to upload their own literary works. It also features groups and instant messaging.

hi5 (http://www.hi5.com)
> hi5 is a popular social network in Latin America, Europe, Asia, and Africa with over 80 million users. hi5 has been released in 26 different languages and is one of the 20 most popular websites, according to Alexa.

Summary

Facebook, MySpace, and LinkedIn are the biggest players in the United States' social networking market. All three sites boast millions of users and allow you to create a user profile and then network with others who have similar hobbies or business/educational backgrounds. Facebook has seen much innovation in the past several years with the launch of Facebook Pages and Facebook Applications, both of which are useful for marketing purposes. MySpace is still a popular social network, but the profound effect of marketing is not as widespread as it may have been in past years. MySpace is great for music marketing, but does not provide many benefits for other marketing niches. LinkedIn is a superior tool for the professional who wants to market himself, while also allowing users to associate themselves with groups and allowing group owners to target members with special messaging and discussions.

There are a number of social networks abroad, including Friendster (which initially gained popularity in the United States but now is more prevalent in Asia), the Brazilian-affiliated Orkut, Bebo, and hi5. If you need to target the overseas community, it is important to join these sites to gain an understanding of their demographics before you proceed with a social media marketing campaign.

Endnotes

1. *http://www.techcrunch.com/2007/03/09/career-advice-dont-choose-facebook-over-your-job*
2. *http://www.facebook.com/press/info.php?statistics*
3. *http://press.linkedin.com/about*
4. *http://www.cartoonbarry.com/2008/01/ive_been_penalized_in_linkedin.html*

CHAPTER EIGHT

Informing Your Public: The Informational Social Networks

There is more to social networking than just connecting users who have the same taste in food, music, and authors. Most of our online interactions are likely communal. We often search for a product on a website and read user reviews. We may try to define a word or find out more about a type of cuisine, and the first result that comes up on the search engine is likely a social site such as a user-generated answer site or even a user-contributed encyclopedia. Indeed, the Web is becoming increasingly social, and we're seeing these interactions in our everyday lives even if we're not cognizant of the fact.

Social media has penetrated all different sorts of online tasks. Collaborative community sites are at the forefront. Community members enjoy helping other community members, and valuable and repeated contributions are rewarded. Therefore, we reach another category of social media that relates to informational social networking sites. Users flock to these sites to share and contribute their knowledge, and in turn, they feel as if they're contributing to the greater good. Of course, there's always some benefit to users who get involved, as those who are active and consistently engaged often receive accolades and are recognized in their communities as valuable contributors. In the end, community members win by getting updated information, and contributors win by being recognized as the people who have provided the community with the answers.

The biggest collaborative informational websites make it easier for users to find answers on the Internet. Whether they are question and answer sites or user-generated encyclopedias, they are popular because they are highly respected and often provide very accurate and detailed historical information not available on other websites.

Human-Edited Social Search

If you perform a search on a country or event, in all likelihood, the search engine will provide you with a result pointing to one of two social sites: Wikipedia or Mahalo. Both of these sites are highly respected—though Wikipeida is a much higher authority—and as such, search engine algorithms place heavy emphasis on them.

Both Wikipedia and Mahalo are human-edited, which means that any user or site contributor can make changes and updates to them. There are thousands of articles on both sites, and new articles are being added daily. Both sites typically strive for accuracy, and those heavily invested on the sites often track any changes made to pages of topics they feel passionately about. Doing so helps ensure factual results.

Wikipedia: The Living Encyclopedia

Launched in 2001, Wikipedia is one of the largest online references on the Internet, with over 12 million articles in more than 260 languages. It is also the top educational and reference website, according to Pew Research.[1] The name "Wikipedia" comes from "wiki," which refers to technology that encourages collaboration (such as initial creation and modification) and "pedia," which refers to encyclopedia. True to its name, Wikipedia has millions of contributors all over the world and hundreds of administrators who actively patrol the site to ensure that there are no abuses and that references are truthful.

Wikipedia is a very open platform. Anyone can contribute, regardless of whether they have an account, but account registration entitles users to have their own profile pages (called the User page) and to establish themselves as credible experts in the community. If you do not have a user account and make modifications to a page, the page history shows your IP address. In the revision history (Figure 8-1), those with usernames are members of the site; the IP addresses listed are associated with users who simply edited the page without logging in.

While Wikipedia is open, it does not come without usage restrictions. One of the most important rules on Wikipedia is that the site will only include articles that cover people, places, and objects of notable importance; notability typically requires "significant coverage in reliable sources that are independent of the subject."[2] Therefore, there are limitations on what people can add to the site, and not every single person or company deserves inclusion in Wikipedia according to its rules and regulations (which are strictly enforced by the site administrators).

Another restriction is that Wikipedia aims to present facts, and nothing but the facts. Entries are supposed to be completely neutral and not opinionated. Any biased commentary is sought out and reedited by Wikipedia administrators and users who strive to publish only unprejudiced entries on the site.

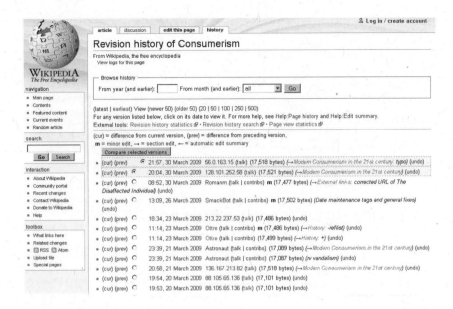

FIGURE 8-1. Wikipedia revision history

Structure of a Wikipedia entry

Every Wikipedia page entry has a public-facing page, but there are numerous other subpages (see Figure 8-2) that include ongoing discussion related to the subject matter and revision history. Wikipedia users can also make use of editorial controls to edit a page, move a page, and watch a page to track updates and changes.

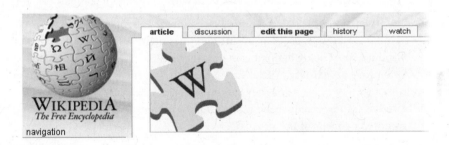

FIGURE 8-2. Wikipedia top navigation shows the various subpages

Article text. When you search Wikipedia, you are presented with the relevant article page (Figure 8-3). The article may be broken down into several different parts; for example, there may be a Contents section that breaks down lengthier articles into subsections, and there may also be a sidebar on the righthand side of the page outlining details about the particular subject matter. For example, if you are viewing an article about a company, the sidebar displays when

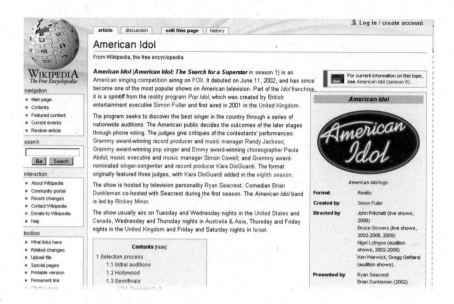

FIGURE 8-3. The main article for a subject on Wikipedia

it was founded and its key executives, or if you are viewing an article about a television show, the sidebar may list the main characters, all of whom likely have their own presence on Wikipedia. Wikipedia articles often include footnotes (called Notes) and External Links on the bottom of the page. The Notes refer to citations relating to the article, and the External Links section features links to sites that are not part of the Wikipedia.org domain but still feature highly relevant content (for example, an actor's official bio page, unofficial but highly referenced pages of movies, and unbiased interviews with founders of companies).

Discussion. The Discussion link is a link to the Discussion or Talk page (Figure 8-4). This page is very important for businesses, as it lets business entities address factual discrepancies themselves, without fear of being banned from the service. The Talk page simply lets users discuss how to improve an existing article. Talk pages can be very simple, with small notes contributed from a variety of users, or they can be extremely exhaustive, with article milestones, biased notes (especially if the subject matter is controversial), and other observations as noted by users of the site.

Both Talk pages and article pages have an "Edit this page" link that allows users to add and remove content. The Wikipedia formatting may be cumbersome, as it is not straightforward HTML, but there are a number of help documents on Wikipedia that give you the stylistic guidelines and formatting conventions necessary for page edits. Wikipedia contains a WYSIWYG editor that features basic functionality; you can take advantage of additional formatting if you have a solid understanding of the Wikipedia formatting guidelines (most of which can be seen at *http://www.en.wikipedia.org/wiki/How_to_edit*). When in doubt on how to format something (especially if you're creating a new page from scratch), you can

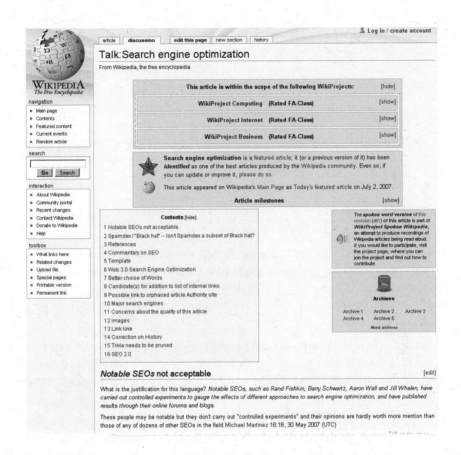

FIGURE 8-4. A Wikipedia discussion page shows contributions by members and lengthy discussions

always go to an existing Wikipedia page, click "Edit this page," and copy/paste the relevant code snippets so that you will have a basic template for your own new Wikipedia entry. You can then preview your updates before publishing the page. Typically, when you make changes, it's advisable to notate the changes you've made by clicking Edit Summary and using the available checkbox to indicate whether it was a minor edit (such as a grammatical or spelling change) or a more major edit, in which case, details of the edit(s) are helpful.

History. If you want to review changes made to any particular Wikipedia article, you can navigate to the History tab to see its revision history. The Revision History page is divided into several columns. The (cur) link refers to the current version of the Wikipedia page, and the (prev) link will let you see what the Wikipedia page looked like before the most recent edit. Typically, you can see exactly what was changed by clicking the (prev) link; Wikipedia is able to pinpoint exactly where the update was made within the context of the entire article and only shows you the pertinent page snippets.

The next column shows you exactly when an update was made, followed by the username or IP address of the person who performed the edit. If the username is blue, that person has a User:Talk page, but if the username is red, there is no existing Talk page for the user. If you click the red link, you will be presented with the standard Wikipedia template to create that User:Talk page (yes, you can edit someone else's Talk page!).

> **NOTE**
>
> The User:Talk page is very similar to the Article talk pages discussed earlier in this chapter. This page corresponds to a user's profile and lets Wikipedia community members converse among themselves and customize their own presence on the site.

An "m" in the next column indicates a "minor edit." The number of bytes of the new Wikipedia entry typically follows, and the Edit Summary link is the last item.

Other Wikipedia features. The final links on the top navigation bar relate to moving pages (the Move tab) and watching updates to pages (the Watch tab). The Move tab allows you to rename an article page; for example, if your company, ACMEWeb, has been renamed to ACMETech, you can use the Move link to specify a new page title. When you do so, you'll be asked for a reason.

The Watch tab is useful if you are using Wikipedia to monitor your business pages or other pages that you find interesting. To utilize this feature, you must have a Wikipedia account. Your watchlist page will notify you when there are edits to articles that you are monitoring.

Wikipedia social media optimization

As established earlier in this chapter, Wikipedia is a website with high authority and is ranked near the top of search results for nearly every search term. This also means that Wikipedia pages get a lot of search traffic, thus explaining most of its popularity. Wikipedia presents a terrific opportunity for you to improve your brand and address reputation management issues. It is also a great way to get a lot of targeted traffic, as popular sites claim that they have seen hundreds of thousands of referral visits from Wikipedia.

Now that you understand the basics of Wikipedia, it's time to discuss how you can contribute to the site in line with both its guidelines and most corporate policies. While Wikipedia does not, at the offset, appear to be a heavily policed website, it certainly has plenty of administrators who ensure that rules are not broken; abuse of the rules results in penalties.

Per Wikipedia policy, individuals who personally have a hand in creating their own Wikipedia pages or employees who use Wikipedia to edit their own company pages can get punished. Wikipedia policy does not allow for conflicts of interest. If Wikipedia does not have an existing page on an individual or company and you believe you can show proof of notability, the best thing to do is to get someone who does not have any conflicts of interest to create the Wikipedia page. Because of the notability rule, you will need to clearly link to multiple sources for the subject to be deemed notable, and you should not be surprised if Wikipedia's administrators feel that the article should be deleted. Once you have created a page, watch it to follow any changes or edits that are made. You can use the Watch link at the top of each page to monitor changes, or you can subscribe to an RSS feed of the page's revision history.

If you are interacting with a Wikipedia page that is a conflict of interest, you are not allowed to make direct changes to the article yourself. Instead, you are encouraged to use the article Talk pages to discuss any factual discrepancies among Wikipedia editors and users, who will then edit the article themselves. Wikipedia also has a Conflict of Interest Noticeboard (*http://www.en.wikipedia.org/wiki/Wikipedia:Conflict_of_interest/Noticeboard*) that you can use to propose changes, especially when the Talk page does not seem to be faring to your advantage or is ignored.

Many company policies won't allow you to edit your company's Wikipedia page at all. Instead, they encourage you to get your feet wet in Wikipedia's social media community by editing pages on topics that you are familiar with and that are unrelated to your organization. When information on a Wikipedia entry is clearly not factual, use the resources available to you (the article Talk page or the Conflict of Interest Noticeboard) to have someone who is in the proper position in your company to communicate on the company's behalf. Many Wikipedia administrators use a tool called WikiScanner (*http://www.wikiscanner.virgil.gr*), which can search a database of millions of Wikipedia edits to determine where edits originated, and can even catch company employees making changes under the radar.

Some Wikipedia optimizers argue that even if you are actually in a position to make changes to your Wikipedia page, you should still create and make changes to Wikipedia pages using a persona and IP address not traceable to your business or company so that the source of the changes will go undetected by the greater Wikipedia community. This has worked well for some contributors on the site. However, Wikipedia administrators are very wary of any subversive tactics to undermine the site's strict guidelines, and they argue that your anonymity on these sites is an illusion that is bound to be exposed. In fact, if you do plan on being promoted to an administrative role on Wikipedia so that you can use the site for your own gain, you will likely be found out and removed from the service; activity of administrators, too, is closely monitored.

Use caution when editing Wikipedia

To avoid being banned from the service, there are other important factors to keep in mind when using it. The following items should be carefully considered when you embark on a social media marketing campaign with Wikipedia.

Wikipedia is not an appropriate place to build links. Wikipedia is not a resource that you can use solely to build links. As stated in Chapter 4, your motives should be altruistic on social media sites. If you are using the site for the sole purpose of adding links at the end of related articles, you will ultimately be found out and likely banned. Instead, build credibility first;

contribute to articles by removing factual discrepancies and fixing grammatical/formatting errors. If you do find that you can add a valuable link as a reference to an article, it's in your best interest to have a list of other altruistic edits to the site before volunteering this information, as adding links too early and often can be perceived as being too self-promotional.

Avoid spamming on Wikipedia: You will get caught. Spam on Wikipedia is frowned upon. If you excessively spam on Wikipedia, you can risk being added to the site's spam blacklist, which serves to highlight users and IP addresses that repeatedly abuse Wikipedia and disregard its rules and policies. Users on the blacklist are prevented from using the site. Search engines and other highly respected websites reference the Wikipedia blacklist, so this is never a good list to have your name or IP address added to.

The contents of Wikipedia belong to the public. Once published to Wikipedia, your company page and even your individual biography do not belong to you. Once that page becomes public, the community owns the content; editors and contributors can add factual content that may be negative and not in the best interest of your company objectives. If your company has been publicly ridiculed and there are references to support the incident, it is fair game for Wikipedia inclusion. You simply won't be able to remove this content. What you can do is aim to find other positive content that is just as factual to balance out the negativity.

Build relationships. If you want to be a regular, respected member within the Wikipedia community and eventually become an administrator, it's important to network with current administrators and establish yourself as a valuable community participant. This entails maintaining activity of your profile and making altruistic rather than self-serving edits. Of course, if you do plan on editing pages that relate directly to your company (or other entries that may be perceived as a conflict of interest), keep in mind that these occasional self-serving edits should be balanced with other altruistic edits.

Mahalo: The Best of the Web

In mid-2007, an eligible Wikipedia contender was launched. Called Mahalo, the site is a human-edited search engine. The service employs quality reviewers and volunteers to make changes to content on its site and to add high-quality and pertinent reference pages. The goal of Mahalo is to have real people sort out the absolute best of the Web and to provide unbiased results that have an advantage over computer-based algorithms. Computer algorithms may not necessarily be correct in deducing whether the results are relevant; with human intervention, however, the absolute "cream of the crop" can be discovered. People can then review and endorse the content. While it is not as popular as Wikipedia, Mahalo results are occasionally ranked at the top of search engine results pages, and therefore, it is an important site to watch and understand.

The structure of Mahalo

Mahalo has several components: Mahalo.com, Mahalo Greenhouse, Mahalo Social, and Mahalo Answers. Each of these components serves a distinct purpose.

Mahalo.com is similar to Wikipedia. When you go to Mahalo.com (Figure 8-5), you can scroll down to the Categories section to find information on thousands of topics. Unlike Wikipedia, Mahalo also provides detailed information on entertainment, such as video game walkthroughs and movie reviews. It also addresses many how-to questions with detailed articles and videos. The structure of a Mahalo page is a bit different from Wikipedia, however; on each Mahalo page, you are presented with the "Mahalo Top 7," which refers to links that Mahalo employees believe provide you with the best user experience and the most relevant information for that specific search query. Depending on the type of page, other content relevant to the topic is presented, and media, such as video, may also be included.

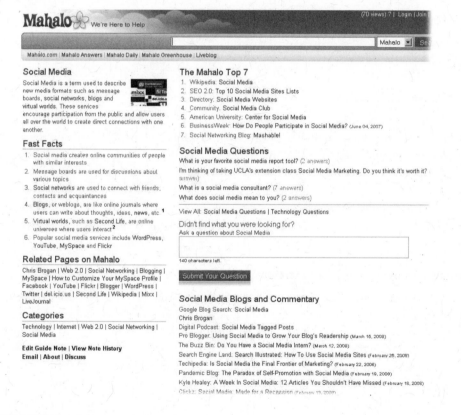

FIGURE 8-5. Mahalo's main page

Links in the Mahalo Top 7 may be accompanied by a warning tag, a hand signal indicating Guide's Choice, or a What Is? designation (Figure 8-6). The warning tag indicates that the link points to a page that is heavy on advertisements such as intrusive banners and pop ups, or that requires users to sign up for a free account. The Guide's Choice image denotes that the link has high-quality and well-designed content; if you mouse over the image, you will see the criteria that Mahalo editors used to determine this designation. The "What Is" image, which looks like a speech bubble with a question mark in it, gives information about a page's credibility, especially when the source of origin is unclear.

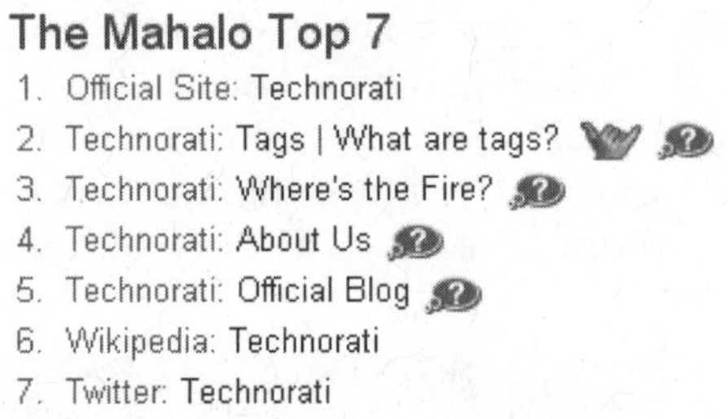

The Mahalo Top 7

1. Official Site: Technorati
2. Technorati: Tags | What are tags?
3. Technorati: Where's the Fire?
4. Technorati: About Us
5. Technorati: Official Blog
6. Wikipedia: Technorati
7. Twitter: Technorati

FIGURE 8-6. Mahalo Top 7 links

Contributing to Mahalo

If you want to become an established community participant on Mahalo, you need to have an account on the site. You can then apply as a Mahalo Guide, which will give you the ability to create Mahalo search pages (also known as search results pages or SeRPs). Mahalo has hundreds of guides working in both full- and part-time capacities. To apply as a guide, you will need to fill out an application at *http://www.mahalo.com/How_to_be_a_Mahalo_Guide*. Qualified guides are paid per accepted SeRP, which gives them incentives to contribute high-quality and well-researched content. Mahalo traditionally seeks applicants with excellent writing, research, and analytical skills.

Mahalo Greenhouse: Curating a search page

Before content is pushed out to the public Mahalo.com website, the SeRPs are built in the Mahalo Greenhouse (Figure 8-7). Part-time guides can contribute their skills at this time, and when a page is complete, full-time guides will review and approve the page. At that point, the page is moved from the Greenhouse, where it was once a work in progress, to Mahalo.com, where it is now available for public consumption. Content created in the Mahalo Greenhouse is not searchable by search engines or users, but content moved to Mahalo.com is.

Of course, like Wikipedia, Mahalo has strict spam guidelines. Only highly relevant websites are appropriate for inclusion in the Mahalo.com pages, and naturally, spam content is frowned upon and will be rejected. Mahalo is also very stringent in the way it handles copyrighted content; sites that are known to violate copyright law are banned from Mahalo and will not be accepted on the site at any time.

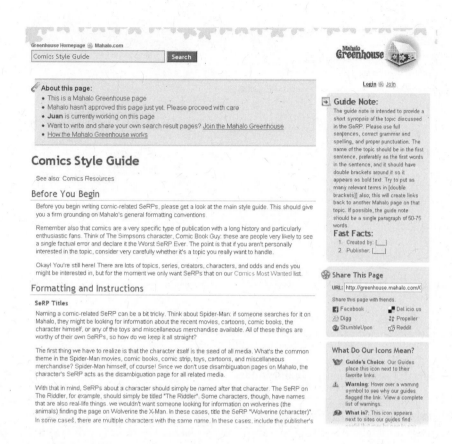

Greenhouse Homepage Mahalo.com

Comics Style Guide   Search

Mahalo
Greenhouse

Login Join

About this page:
- This is a Mahalo Greenhouse page
- Mahalo hasn't approved this page just yet. Please proceed with care
- **Juan** is currently working on this page
- Want to write and share your own search result pages? Join the Mahalo Greenhouse
- How the Mahalo Greenhouse works

Guide Note:
The guide note is intended to provide a short synopsis of the topic discussed in the SeRP. Please use full sentences, correct grammar and spelling, and proper punctuation. The name of the topic should be in the first sentence, preferably as the first words in the sentence, and it should have double brackets around it so it appears as bold text. Try to put as many relevant terms in [double brackets]] also; this will create links back to another Mahalo page on that topic. If possible, the guide note should be a single paragraph of 50-75 words.

Comics Style Guide

See also: Comics Resources

Before You Begin

Before you begin writing comic-related SeRPs, please get a look at the main style guide. This should give you a firm grounding on Mahalo's general formatting conventions.

Remember also that comics are a very specific type of publication with a long history and particularly enthusiastic fans. Think of The Simpsons character, Comic Book Guy: these are people very likely to see a single factual error and declare it the Worst SeRP Ever. The point is that if you aren't personally interested in the topic, consider very carefully whether it's a topic you really want to handle.

Okay! You're still here! There are lots of topics, series, creators, characters, and odds and ends you might be interested in, but for the moment we only want SeRPs that on our Comics Most Wanted list.

Formatting and Instructions

SeRP Titles

Naming a comic-related SeRP can be a bit tricky. Think about Spider-Man: if someone searches for it on Mahalo, they might be looking for information about the recent movies, cartoons, comic books, the character himself, or any of the toys and miscellaneous merchandise available. All of these things are worthy of their own SeRPs, so how do we keep it all straight?

The first thing we have to realize is that the character itself is the seed of all media. What's the common theme in the Spider-Man movies, comic books, comic strip, toys, cartoons, and miscellaneous merchandise? Spider-Man himself, of course! Since we don't use disambiguation pages on Mahalo, the character's SeRP acts as the disambiguation page for all related media.

With that in mind, SeRPs about a character should simply be named after that character. The SeRP on The Riddler, for example, should simply be titled "The Riddler". Some characters, though, have names that are also real-life things. we wouldn't want someone looking for information on wolverines (the animals) finding the page on Wolverine the X-Man. In these cases, title the SeRP "Wolverine (character)". In some cases, there are multiple characters with the same name. In these cases, include the publisher's

Fast Facts:
1. Created by: [___]
2. Publisher: [___]

Share This Page
URL: http://greenhouse.mahalo.com/(

Share this page with friends.
 Facebook   Del.icio.us
 Digg   Propeller
 StumbleUpon   Reddit

What Do Our Icons Mean?
 Guide's Choice: Our Guides place this icon next to their favorite links.
 Warning: Hover over a warning symbol to see why our guides flagged the link. View a complete list of warnings.
 What is?: This icon appears next to sites our guides find

FIGURE 8-7. A Mahalo Greenhouse page is a work in progress until it is moved to Mahalo's main site

If you are not a Mahalo guide, you can still contribute to the site on a lesser scale. You will be unable to create pages from scratch, but if you have an account on Mahalo.com, you can make edits to individual pages. Unlike Wikipedia pages, these pages will be confirmed by Mahalo's full-time staff for accuracy.

Further, Mahalo does let you edit pages about yourself or your company.[3] Mahalo's full-time staff members will still check all edits for accuracy and approve or reject the changes, but contrary to Wikipedia rules, writing about yourself is not frowned upon. However, some corporate social media policies will still prohibit employees from making edits on behalf of the company and will instead elect a social media spokesperson to handle these communications.

On every page on Mahalo.com, a Discuss link (typically found at the bottom of the entry) points users to a message board where they can converse with one another and with Mahalo Guides about specific content of the page (Figure 8-8). The message board allows site members to discuss the search result page, the links contained therein, and the overall topic.

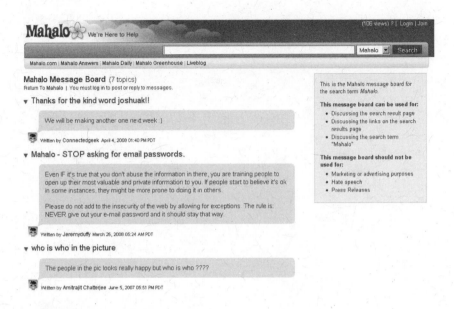

FIGURE 8-8. Mahalo discussion boards let site members discuss content of the current entry

Mahalo Social: The community of contributors

There is a lot more to Mahalo than its basic search engine. Mahalo Social (Figure 8-9) is the social networking arm of Mahalo and allows Mahalo members to connect with one another. For the most part, Mahalo Social simply aggregates social media profiles, such as your MySpace profile, your delicious.com account, and your Twitter profile, among any other social media profiles you maintain. More importantly, Mahalo Social enables users to recommend links to the site and to one another. These links can then be used to further enhance the search results on Mahalo.com.

The links that you recommend on Mahalo still must go through editorial review and can be rejected or banned.[4] Mahalo will typically only allow the addition of links that are from highly respected or heavily trafficked websites. This policy stems from Mahalo's desire to give users the most relevant content. If the content is laden with spam or if the site is not well developed (and there is simply not enough content to justify its addition to any page), editors may ignore the recommended link. Mahalo allows you to track the submissions of your recommended links to determine whether they have been accepted or ignored. The service also incentivizes by ranking your profile based on the number of successful links added to the site.

The Mahalo Social service also allows you to network with other users who maintain Mahalo accounts. You can network with friends to see their recommended links and other account activity on Mahalo. While there is no easy way to directly and privately communicate with

FIGURE 8-9. A Mahalo Social profile page shows activity across the site, including links and Mahalo Answers activity

these users on Mahalo.com, if an account holder has shared her other social media profiles on Mahalo Social, you can communicate with her on those other social media channels.

Mahalo Social has a toolbar called Mahalo Follow (Figure 8-10), which allows you to quickly recommend links to Mahalo.com, to share your favorite links with others via other social media channels, to discover your friends' favorite content as shared on Mahalo, and to easily search within existing Mahalo pages. The toolbar also incorporates Mahalo results into your typical search results so that you can easily find related Mahalo material from a regular search. You can download the Mahalo Follow toolbar at *http://www.greenhouse.mahalo.com/Special: FollowDownload.*

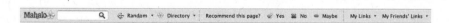

FIGURE 8-10. The Mahalo Follow toolbar gives you special access

Mahalo Answers: Ask and ye shall find

One of the latest additions to Mahalo is Mahalo Answers (Figure 8-11), which is a knowledge exchange that allows users of the community to ask questions and provide answers to the questions submitted by other users. The entire site is community-driven: any member of the site can participate by asking any question she wants. Other community members can then provide answers. Answers can be ranked by any user on the service as "helpful" or "not helpful," but the asker chooses the best answer.

FIGURE 8-11. Mahalo Answers lets members help other members

Perhaps most important to Mahalo Answers is the ability to "tip" the best answerer with real monetary currency. When you ask a question of the Mahalo Answers community, you typically put a Mahalo Dollars monetary value on each individual question. This can help ensure that your question gets prompt attention. If you find an answer that is very helpful, you can then "tip" the answerer with the amount at which you valued your question. Each

Mahalo Dollar can be cashed out at $0.75 per $1 once a user's account balance exceeds $40 (Mahalo monetizes its service by taking $0.25 per every $1).

If you choose not to tip any answerer, you can ask for a refund of that money and rescind the tip, but be advised that this action is reflected in your user profile. To build trust in the Mahalo Answers community, you should deliver on your monetary promise, especially if the answer provided by another community member is valuable and helpful.

Mahalo Answers may be syndicated to actual Mahalo.com search engine articles. This also gives you additional exposure as a contributor to the site.

Choosing the Best Answers: Using Yahoo! Answers for Social Media Marketing

The Internet is social by nature, and as such, question-and-answer knowledge exchanges are here to stay. People enjoy soliciting the advice of their peers on just about every subject to get the benefit of others' firsthand experience. Answers are especially helpful because unlike search results, they are biased; humans can specifically recommend products to solve particular problems, or they can recount experiences that will empower other community members in similar situations. The most important social networks for this purpose boast millions of monthly visitors and are also highly ranked in search results pages. Therefore, they are great tools for social media marketing and can be used very effectively when you understand the dynamics of the sites and how to leverage them appropriately.

The most popular Q&A site is Yahoo! Answers (*http://www.answers.yahoo.com*), shown in Figure 8-12. In a 2008 Hitwise survey,[5] it was found that nearly 75% of all visits to question-and-answer websites were being directed to Yahoo! Answers, making it the most authoritative reference site for questions. Yahoo! Answers has hundreds of millions of regular users, and there are hundreds of millions of questions and possibly billions of answers provided on the network. If you've ever typed a question in a search engine, you've likely been presented with a link to a Yahoo! Answers page that directly addressed your inquiry. Yahoo! Answers can, therefore, be a very powerful social media marketing tool, as it can drive traffic and awareness. With an understanding of Yahoo! Answers, you will be able to use this service to your advantage.

Yahoo! Answers works much like Mahalo Answers. Participants ask questions, and other community members provide answers. Instead of a monetary incentive, Yahoo! Answers uses a point-based system. Community members accumulate points for answers they provide; answers that are voted "best answer" receive more points. (Yahoo! will deduct points from users who ask questions.) If you are able to answer question and achieve a high point ranking, you can establish a good reputation on the site. The higher ranks also give you privileges not offered to users in lower rankings, such as the ability to ask more questions in a specified time period and to vote on answers more frequently.

FIGURE 8-12. Yahoo! Answers home page

There's an abundance of activity on the Yahoo! Answers site. At any given time, questions are being asked and answers are being provided. This means that if you plan on using Yahoo! Answers to provide your own solutions to the community, you have an almost unlimited chance to do so, as there are hundreds of related questions that you likely can contribute your answer to.

Yahoo! Answers is one of the busiest web properties, with community engagement during all hours of the day. There are millions of archived questions and hundreds of thousands of questions that you can contribute to as a community member.

Promote Your Business on Yahoo! Answers

Your business likely provides a service or product. Maybe you have a web page that directly answers a question that has been asked on Yahoo! Answers. Wouldn't you like to capitalize on the opportunity to give a valuable answer to someone who may be seeking out your expertise?

Fortunately, you can. However, you need to make sure you follow the appropriate rules of engagement. Like Wikipedia, Yahoo! Answers relies heavily on community moderation. If you are perceived to be submitting spam, users can click the Report Abuse button and write an explanation of why a particular question or answer may not be appropriate for the Yahoo! Answers community; the user account, question, or answer may subsequently be deleted.

Give Back to the Community

Like other social networks, your motives should still be somewhat altruistic in nature, though later in this section, you will learn that fulfilling business objectives is not a punishable tactic. Since Yahoo! Answers has so many questions, there should be no shortage of questions you can answer, and further, you may notice that some of the answers provided by regular community participants are not necessarily the best and most comprehensive. This is where you have the opportunity to shine and really add value to the community. By writing well-researched and helpful answers, you can establish yourself as a highly ranked member on the service. The more "Best Answers" you accumulate on your account, the more opportunity you have to build this credibility.

Once you find a question you feel you can provide a great answer to, click the Answer this Question button. You will be presented with a screen that lets you compose your answer (Figure 8-13).

Of particular note is the field below the Your Answer box. Use this field to specify your source. That source may be Wikipedia, a news article, or even a web page you wrote yourself on your own company's website! This is a great opportunity to provide basic information that answers the question and refers readers to more detailed information.

Find Questions to Answer

If you want to take advantage of the Yahoo! Answers facility on a semiconsistent or regular basis for marketing, you can refer to the Advanced Answer Search to find searches for a particular phrase. Ensure that you indicate that you are only looking to search for Open Questions; this way, you will be able to participate in any relevant Q&A thread that has not been closed due to expired time. Naturally, if you proceed along this route, you may wish to balance these marketing-type contributions with your own altruistic edits; chances are you will be able to answer an 8th grader's homework question, or perhaps even pregnancy

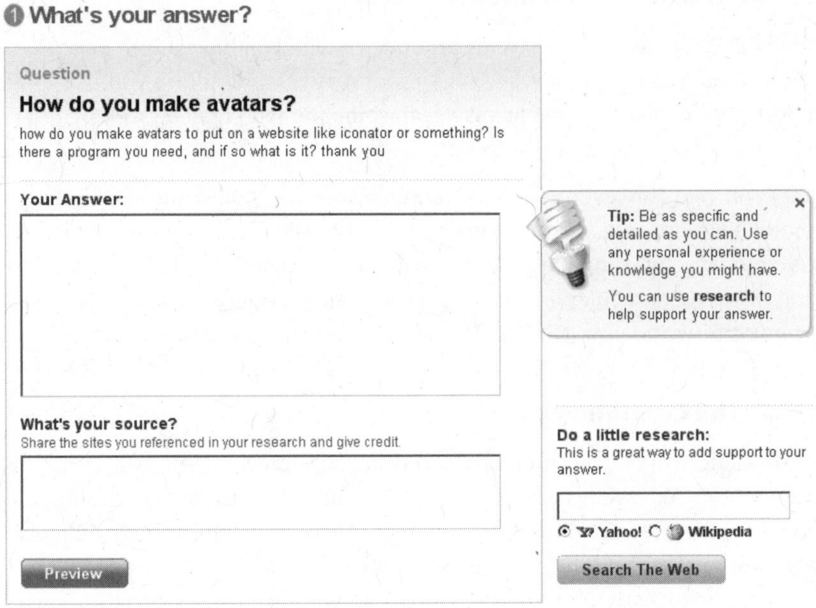

① What's your answer?

FIGURE 8-13. Providing an answer

questions. Yahoo! Answers has a plethora of questions pertaining to homework and pregnancy; these are two of the most popular and active categories on the site.

Associate Yourself with Your Business

If you do not have any specific business objectives but perhaps have expertise in a particular subject matter, you may want to leverage Yahoo! Answers in a different way. You can add links to your Yahoo! Answers profile in the "About Me" section, so this is a good place to showcase your company website or blog. You can also sign your answers with an associated URL, especially if you feel that it supports your answer and your credibility. It is never a bad thing to identify yourself as a bartender when asked about liquor, and it's certainly not harmful to identify yourself as a veterinarian if you are answering a question about the health of someone's cat. Of course, signing your answer with your name, affiliation, and business home page may be a good way to get more eyeballs on the business you provide.

You're Encouraged to Self-Promote!

The best part about Yahoo! Answers is that none of this self-promoting activity is frowned upon by Yahoo!. In fact, this behavior is encouraged. According to the site's own guidelines, you should associate yourself as a knowledgeable professional or business owner when appropriate:[6]

There are many professionals and business owners providing valuable knowledge and experiences on Answers. By identifying yourself as such and providing great answers, you are building credibility and positive brand image.

By all means, then, use Yahoo! Answers as often as possible for your business marketing objectives. As you contribute more and more to the site and your answers are voted "Best Answer" on a regular basis, you'll have more of an opportunity to really stand out from the crowd. Perhaps that activity can get you voted as "Top Contributor" on the site, which is a designation provided by Yahoo! Answers to active participants in particular categories.

Keep Abreast of the Latest Questions, and Answer Away

So how do you monitor Yahoo! Answers on a regular basis? You can visit the site frequently and navigate to a specific category, or you can perform a search. Better yet, you can subscribe to RSS results of a particular search on Yahoo! Answers, so those earlier Advanced Search results can be ported directly into your RSS reader of choice. Similarly, you can also subscribe to RSS feeds of any general category, as featured on the left-hand sidebar of the Yahoo! Answers site. Once you get these questions through your feed reader or via the search, you can begin answering more and more relevant questions and building your account persona to establish trustworthiness on the site.

Other Q&A Websites to Be Considered in a Social Media Marketing Strategy

While Yahoo! Answers has the largest share of visits among all knowledge exchange websites, there are numerous other websites that are very similar in approach and can also work to your marketing advantage. Be advised that when you engage on such sites, you should avoid any excessive self-promotion; some sites will not approve of marketing at all.

WikiAnswers (http://www.wiki.answers.com)
WikiAnswers is the second most popular Q&A website, according to Hitwise, with approximately 19% of total visits. The service is a hybrid of the most interesting features of Yahoo! Answers and Wikipedia. Like Yahoo! Answers, users ask questions, and members of the community answer the questions. And, like Wikipedia, you can enhance existing answers and watch updates. Therefore, only one answer is provided and can be edited at any given time by WikiAnswers account holders for accuracy.

Ask Metafilter (http://www.ask.metafilter.com)
Ask Metafilter is a very close-knit community of individuals interested in helping one another. Answers are typically higher quality than many other knowledge exchange websites, and community members are extremely engaged and typically passionate about the subject of discussion. As such, Ask Metafilter strives only to accept users who are serious contributors; each new signup costs $5. On a site like Ask Metafilter, it is critical

to avoid being overtly self-promotional, as that can result in a lifetime ban from the service (and the $5 you initially donated is nonrefundable). The best contributors are those who are selfless and really want to enhance the community by providing helpful answers. The community is cognizant that there are numerous users of the service using it for marketing gain. Therefore, you should avoid contributing in a self-centered way.

Answerbag (http://www.answerbag.com)

Answerbag is very similar to Yahoo! Answers, as it gives users the ability to ask questions and answer them. Users can rate questions and answers, and perhaps most significantly, users can comment on answers. If you've seen some of the answers provided on Yahoo! Answers, you know that having the ability to comment on answers is a useful resource, especially when the proposed answer is completely off-base (and yet is still somehow endorsed by the community as the most relevant answer!).

Askville (http://askville.amazon.com)

Askville, owned by Amazon.com, is a resource-intensive knowledge exchange. Some of the more popular answers on the site are extremely well researched and detailed. Askville encourages you to add citations and references to your answer and also allows you to accompany your answer with a map, a product that can be purchased via Amazon.com, or video (which is pulled by Google Video, YouTube, and other services).

Twitter Answers (http://www.mosio.com/twitter)

Twitter Answers provides a unique way to use Twitter for answering questions. You must have an account on Twitter and also on Mosio.com, a mobile-powered search provider. With Twitter questions, you follow Twitter user "QNA" to ask a question. You can receive up to four answers to your question, which will be posted to the website or directly to your mobile phone. This service is mostly informational and is not necessarily the best option for social media marketing, as answers are provided to one individual (the person who asked the question).

@answerme (http://www.twitter.com/answerme)

@answerme is another Twitter-powered tool for asking and answering questions. Simply start your question with @answerme. Start an answer with @answerme @username and proceed to offer an answer. Again, this may not be the ideal tool for social media marketing, but it may be useful for discovering information and also for targeting your answer to someone specifically seeking out input. Then again, using Twitter as a standalone Q&A tool may be the easiest way to get answers quickly.

LinkedIn Answers (http://www.linkedin.com/answers)

You can use LinkedIn Answers to solicit information from experts in a particular subject. It also provides a great way to connect with individuals within your LinkedIn network and to possibly even gain new customers. For more information on LinkedIn Answers and how to use the service for your marketing goals, see Chapter 7.

Aardvark (http://www.vark.com)

Aardvark was launched in March 2009 and allows you to get answers to questions from users in your extended social network. Aardvark is primarily powered over IM applications. When you ask a question on the Aardvark service via IM, the service tries to farm your contacts and your second-degree contacts (your friends' contacts) to find the best answer to your question. It's not the best tool for marketing, but it shows the power of leveraging your social communities for knowledge gain.

Knowledge Is Power

If you possess knowledge, by all means, don't be afraid to share it. Numerous individuals, bloggers particularly, worry about giving out so much free information on their websites and wonder if it will be detrimental to business. Chances are that you'll be hired regardless of the content you provide on your website or in social profiles, because you'll perceived as someone who can execute and who possesses even more knowledge than you share.

The biggest benefit to sharing knowledge is that you're arming individuals with information that enhances their opinion of your credibility. Would the CEO of a company rather be known as someone who has a great deal of knowledge and appears to be extremely approachable, or would he want to be perceived as someone who has kept all the "secrets" to himself? In today's modern social media marketing mindset, knowledge truly is power and both parties can benefit from it.

What does this ultimately do for you? By contributing knowledge for free to the greater community, you can garner links that point to your social media profiles, your valuable contribution, or to your website. After all, if you provide helpful insights, people will notice, and those who maintain their own social profiles may recommend a link—your link—to their networks. They may reference your incredibly helpful article. They may recommend your company for related business tasks. This is the foundation upon which social media works: you share, and the community shares as well. It starts on knowledge exchange websites and informational sites such as Wikipedia, but the benefits can be much farther-reaching than previously imagined.

The more links you get, the more of an authority you become. This works both for human readers and search engines. If I am finding "endorsements" to your content within a news article and then later read another endorsement on another highly respected social media marketing channel, I am more likely to trust you. Similarly, when search engines see different links from diverse sources pointing to your page, they, too, begin to trust you as a contributor. This is how sites like Wikipedia, Mahalo, and Yahoo! Answers have come to be such highly regarded websites: the knowledge contained therein has been consistently valuable to thousands of users and content creators on the Web.

With links comes traffic. Wikipedia, Mahalo, and Yahoo! Answers are extremely popular websites because they are highly ranked in search engine results pages. Further, good content that is shared among individuals translates to more traffic as well. By taking advantage of these opportunities, you can drive relevant traffic to your website whenever you deem appropriate.

As a result of links and traffic, you can brand yourself and establish your identity as a true thought leader in your industry. If you consistently engage in these communities and provide useful information that the community appreciates, you can become well known within your industry, which ultimately can improve your brand and establish your authority in a specific niche.

Create Your Own Wiki

While you can contribute your expertise to existing websites, you can also create your own wiki by utilizing online services or from freely available open source applications. Wikipedia, for example, is powered by an open source platform known as MediaWiki.

A *wiki* is simply a collection of web pages that can be modified by anyone with access to the web application. It is a living document that allows for collaboration and regular updates.

If you have your own web hosting provider, MediaWiki offers you the familiar Wikipedia look and feel, though newcomers may find that building a wiki from scratch is overwhelming and will still need to understand the syntax for updating and editing wiki pages. If you do not have your own hosting provider or want to host your wiki page elsewhere, consider using pbwiki (*http://www.pbwiki.com*), a widely used collaborative wiki for businesses, college course work, and other user-generated content. Figure 8-14 shows an example pbwiki page for a college course on social networking. pbwiki is available in a free version, but paid options are also available with more storage and support options.

A second hosted option for your wiki is Wetpaint (*http://www.wetpaint.com*), which also offers free and paid options. One of the premier social media marketing wikis, the Altimeter Wiki, is powered by Wetpaint.

A wiki is a great tool to keep the public informed of the latest additions and updates about a particular subject without having to ask a specific forum administrator or blogger to make an update. Further, wikis are extremely informative.

Wikis are best used when targeting a very specific audience that is already active and engaged. Additionally, it is best to have a moderator who can oversee changes on a regular basis.

> TIP
>
> While wikis are powerful applications for collaborative work, they can also be excellent low-cost options to convey internal company information for small businesses. Executives can communicate business updates to the company's employees via a wiki without much overhead and technical know-how.

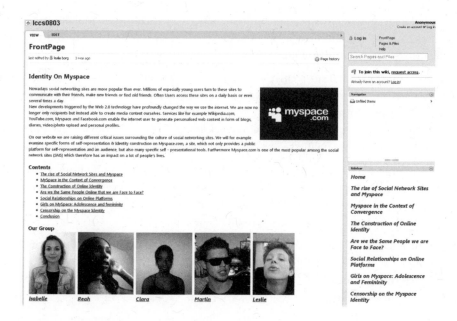

FIGURE 8-14. A pbwiki page serves as a collaborative effort for a college course

Summary

Our interactions on the Internet are likely more social than we ever previously believed possible. Today, we perform searches on the Internet, and social informational sites are likely to be ranked high in search results. Sites like Wikipedia, Mahalo, and Yahoo! Answers are endorsed by users via links and word-of-mouth marketing, and search engines, therefore, know to provide emphasis on these pages in the results. Because community members can contribute to these pages, they provide a great way to build relevant traffic for your brand or company web page. . Furthermore, they help you establish yourself as a knowledgeable expert while building strong relationships in another social media capacity.

Wikipedia is by far the largest social informational network and has indexed millions of articles. When you're contributing to Wikipedia, the most important rules to remember are that spamming is prohibited and that you should only add content that is deemed "notable" by the community. You will need to establish notability for any new page created. Further, Wikipedia does not allow you to edit pages where there could be a perceived conflict of interest, so companies and individuals are not allowed to edit pages about themselves. Instead, employees of a company can use the article Talk pages to voice concerns with page editors, or they can use the Noticeboard to bring the issue to the attention of Wikipedia administrators.

Mahalo is a new human-edited search engine with several components. Mahalo.com content is typically curated in the Mahalo Greenhouse. Guides create SeRPs here and then push them out to the public. Anyone can apply to be a guide on Mahalo. The site also has a social element called Mahalo Social, where you can recommend links to be included on Mahalo.com. Finally, Mahalo has a questions-and-answers section, and the service entices users to contribute valuable answers by offering monetary compensation.

Beyond these informational sites, Q&A websites are also becoming prevalent. By far, the biggest Q&A website is Yahoo! Answers, with millions of members and billions of answers. The best part about Yahoo! Answers is that there are no rules restricting businesses from participating in a self-promotional way. Yahoo! acknowledges that if you do have the business knowledge, you are more qualified to answer some questions than the average layman.

A number of other question-and-answer sites are noteworthy, but not all bring great traffic. Of particular significance is WikiAnswers, which lets people edit answers (as on any other wiki), and Amazon's Askville, which lets users add detailed answers and encourages them to provide relevant widgets, such as maps, product pages, and video.

Being a contributor to these websites can result in links that may work to your benefit by bringing you more traffic, more credibility, and better branding. Ultimately, you can become an established thought leader and perhaps someone others trust or rely on for accurate and helpful contributions or answers.

Anyone can create a wiki on any subject. Wikis allow users to collaborate on particular issues and can be extremely useful for compiling data about a particular topic, especially when new information is always coming in. Some very popular collaborative social sites operate upon wikis, and wikis can also serve as wonderful tools for internal small-business communications.

Endnotes

1. *http://www.pewresearch.org/pubs/460/wikipedia*

2. *http://www.en.wikipedia.org/wiki/Wikipedia:Notability*

3. *http://www.calacanis.com/2008/05/31/wikipedia-3-0-you-can-now-edit-any-page-on-mahalo*

4. *http://www.blog.mahalo.com/2007/06/20/case-study-how-mahalo-handles-user-recommended -links*

5. *http://www.hitwise.com/press-center/hitwiseHS2004/question-and-answer-websites.php*

6. *http://www.help.yahoo.com/l/us/yahoo/answers/abuse/guidelines-10.html*

CHAPTER NINE

Leaving Your Mark: How to Rock the Social Bookmarking Space

You find a website that you like and decide to save it to your Favorites. The next time you're online, you can open your browser, navigate to your Favorites menu, and start accessing the pages you need easily. That's the bookmarking most of us are familiar with. Social bookmarking takes that a step further by allowing users to save their bookmarks online to share with friends. As a plus, your bookmarks are accessible wherever you go and from any computer.

With social bookmarking, Internet users can store and organize their bookmarks publicly. Many applications take this functionality and extend it by allowing users to recommend bookmarks to friends and to label these bookmarks with certain identifiers, called *tags*. This mostly publicly available information can then be used for content discovery and understanding how others categorize content produced by web developers and companies.

A Timeline: The Past, Present, and Future

Bookmarking has gone through a pronounced evolution since its usefulness was first discovered. Whereas bookmarking was once a task for private use, it has since become increasingly social, and the future will likely see even greater integration of bookmarking in our everyday habits and web searches.

The Past: Bookmarking Without the Help of Social Sites

Once upon a time, you discovered a web page you knew you'd want to reference again and again. Chances are that if you wanted to keep tabs on that page without having to memorize the URL, you'd navigate to the Bookmarks or Favorites menu on your web browser and save the website address in a process known as *bookmarking*. Much like the traditional ways of saving your place in a book or manuscript, an Internet bookmark is a pointer to a stored web page location so that you can easily access this information again in the future.

When bookmarking URLs the "old-fashioned" way, you'd likely also categorize your bookmarks into a folder or give them easy-to-remember titles. That way, you could find the page as needed despite the amount of bookmarks you may have accumulated over a lengthy period of time.

But bookmarking has its problems, which have become obvious over time.

Eventually, you'd probably collect many different bookmarks and your browser would be slow to run. After all, if you had gathered hundreds of different bookmarks, your browser would be feeling bloated with all the new personalized data it needed to load before it actually pulled up your home page. Ever wonder why your browser is slow? Chances are, there is simply a lot of personalized data contributing bogging it down!

What about organization of the bookmarks? If you merely saved a bunch of pages in bulk to a folder, how could you possibly find that article you bookmarked six months ago among the clutter in your Favorites menu? If you're a bookmarking packrat, you may have had problems finding a page you stored long ago, and the length of time it took to pull up your Favorites menu probably did not add to the fun of searching through your bookmarks for the single one you were looking for.

These are obstacles that present-day social bookmarking has attempted to solve.

The Present: Socially Shared

With the rise of content production and the cheap cost of establishing a presence on a website, it is now easier for users to sign up for free accounts on websites to store data they really need to reference without having that information confined to a single computer's bloated browser. When one of my supervisors had to have her computer formatted due to a virus, she was worried about leaving her important bookmarks behind. Before the computer was completely wiped out and the operating system reinstalled, we signed her up with an online social bookmarking service where she stored her needed URLs. From that point on, she could simply go to her personalized page to get her favorite bookmarks whenever she needed. The best part for her is the ability to access this data from her home computer and from her work computer, so she isn't restricted to any specific location when she wants to pull up the web pages she needs.

Little did my supervisor know when she signed up on the social bookmarking site that it is jam-packed with features beyond just storing your must-have URLs in a centralized online location. There is a lot more to social bookmarking than remote access from a single location. These services don't just store your bookmarks for your own consumption; everyone can benefit from seeing your bookmarks and favorite URLs. Of course, there's always the concern that your private data could also be discovered via these public channels, but most bookmarking services give you the option to either share these pages publicly to the entire social network or to keep specific URLs private.

When you think about it, how many bookmarks do you have that are completely private—that you would not want to share with anyone? What about public and interesting content? Chances are, you have a few URLs to your bank accounts and perhaps other sensitive information stored somewhere on your computer. On the other hand, you probably have sent your friends, colleagues, and family members hundreds of URLs to other websites because the content on them was interesting. Or you may have stumbled upon an article that is relevant to your interests that you may want to reference later—just like a bookmark. These URLs don't need to be privately saved. In fact, while social bookmarking services can save URLs that you designate as public or private (and yes, you can make all your URLs private if you wish), you'll find that there's a lot to gain by sharing, from building your network to discovering new and attention-grabbing content.

This is another reason that socially shared bookmark applications have taken off in popularity. With hundreds of millions of bookmarks stored on a variety of services, some popular and others still gaining public adoption, it is clear that individuals really like to tell people about their favorite destinations on the Internet. Further, social bookmarking serves the purpose of assisting in research. By tagging your bookmarks, you give semantic value to your bookmarks so that you can easily find them again. But if you tag them *and* share them, individuals in your network or elsewhere can discover brand-new content that can further enhance their work needs or interests. Finally, you can look at the way other people are bookmarking your site to get a feel for public opinion about your organization.

DEFINITION

Tags are metadata typically assigned to a unit of information (in this case, a bookmark) as descriptors. For example, an article about television series *House* may be tagged as "drama," "FOX," "television," "tv," "medicine," "health," or "hughlaurie" (yes, normally without spaces, as will be described later in this chapter), among other potential tags.

Let's look into social bookmarking in more detail. While social bookmarking services exist to help you centralize a list of your favorite websites in a single location, the goal really is for social sharing, and that's exactly what today's social bookmarking accomplishes. The home pages of many widely used social bookmarking sites specifically display the most popular bookmarks of the hour, and this data is constantly being refreshed with brand-new, remarkable and timely content.

How do you actually start saving bookmarks? Depending on the service, there are multiple ways to work with social bookmarking tools. Some services have *bookmarklets*, which refer to snippets of code that are embedded in the browser and act as shortcuts to perform certain tasks. Some have actual add-ons, such as toolbars, for browsers. Some will even allow you to navigate directly to a website and input the bookmark that way. With these tools, you will usually have the opportunity to add tags and categories where applicable.

Once you've actually stored your bookmarks, you can always navigate back to your customized home page from any computer, though you will only be able to access private bookmarks if you're logged in. Typically, social bookmarking sites will also allow you to navigate directly to a tag page to see all bookmarks with a specific tag. This lets you access all of your topical information in one fell swoop.

If you want to access the bookmarks of friends, you can add them to your network seamlessly. Some services will let you add your friends immediately and access all of their information; others will only provide access once a friendship connection has been confirmed.

If you like the bookmarks of friends, you can subscribe to their favorites via RSS. This lets you keep abreast of their newest discoveries as they add new bookmarks to the services. You can also share your favorite links with your friends on these social bookmarking sites, through the process to do so varies on each individual site.

Later in this chapter, we'll discuss the best features of the most popular social bookmarking services to see how you can maximize them for bookmarking and inspiration.

The Future: A Likely Integration into Search Results

Tagging is not just a social bookmarking phenomenon. It facilitates people keeping track of content online and locating those pages later. Gone are the days of purely algorithmically based tasks; today, social elements permeate all of our online activities. Even search engines, which traditionally have relied so heavily on algorithms to provide results, are taking more social cues and integrating human behavior into their search algorithms. Google has been the frontrunner in social integration, as described in an article by search guru Danny Sullivan on search industry premier blog Search Engine Land.[1] In the article, Sullivan suggests that Google will be getting more personal. With the 2007 launch of Google Personalized Search,[2] which provides more personalized results to the web searcher, it's clear that Google is looking to make searching a lot more customized for searchers.

With Google's launch of SearchWiki in November 2008,[3] it certainly appears as if it is pivoting toward the social. With SearchWiki, personalization also plays a tremendous role: users can rerank or completely remove search results. It also allows users to add notes to search results (which seems awfully similar to the tagging we see today among social bookmarking services, and also appears to emulate how one can add descriptions to new bookmarks on social bookmarking sites). While the social role that SearchWiki will play in the future is uncertain, it is possible, according to Google, that user interaction will influence how Google reranks results. [4]

Perhaps we will see the evolution of social bookmarking really make its way into search. Only time will tell.

Using Social Bookmarking Sites

There are two main contenders in the social bookmarking sphere, delicious.com and StumbleUpon. While StumbleUpon is arguably not a bookmarking service at its core—instead, it's a content discovery engine—it contains many of the same elements seen in social bookmarking services, and since it can be and often is used as a bookmarking tool, we will explore it further in this chapter. Yahoo!-owned delicious.com, on the other hand, is by far the leading social bookmarking service on the Internet. We will look at how to use these services for storing bookmarks, retrieving bookmarks, finding new content, and networking.

A third site, Diigo, is becoming increasingly popular for social bookmarking. Like delicious.com, it is a social bookmarking tool but with a lot more feature enhancements that are not available in the widespread delicious.com application.

StumbleUpon: A Content Discovery Engine with Bookmarking Features

One of the most unique social media offerings is StumbleUpon, a social content discovery engine with bookmarking features. StumbleUpon is different from many other social sites in that it works via a toolbar installation on your browser. Once it gathers personalized information from you (hobbies and interests), you can start surfing with StumbleUpon to find brand new sites that are related to your interests as suggested by other users on the service. The more active you are on StumbleUpon, the more opportunity there is for you to grow your network and expose your own content to more and more StumbleUpon users.

Signing up with StumbleUpon

If you're ready to take the plunge into StumbleUpon, you must have a supported web browser. StumbleUpon supports both Internet Explorer and Mozilla Firefox. The first step is to choose a username, which will ultimately point to your personalized space on the site, such as http:// yourname.stumbleupon.com. You will then be asked to download the software. After downloading and installing the software, provide information about your StumbleUpon community goals (are you here for business? friendship? web surfing? dating?). You can also volunteer other information, such as your interests, favorite TV shows and movies, and more. Next it's time to manage your interests (Figure 9-1). In this section, mark the checkbox of the categories you

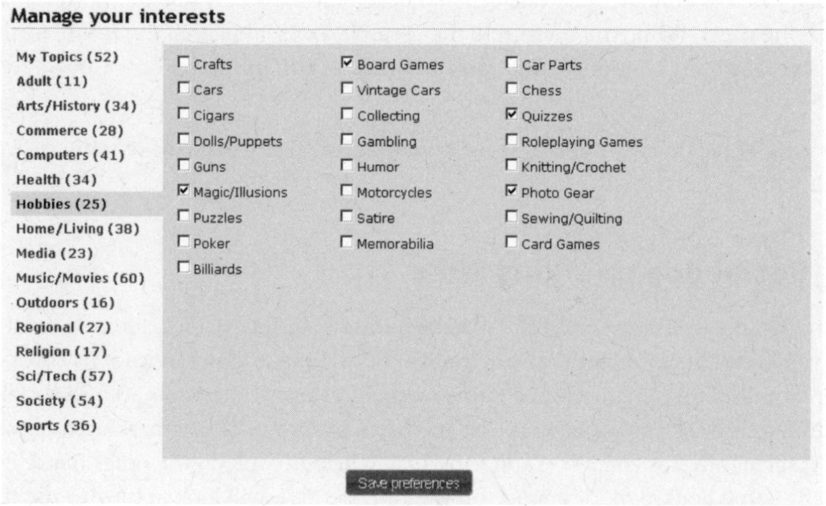

Manage your interests
My Topics (52)
Adult (11)
Arts/History (34)
Commerce (28)
Computers (41)
Health (34)
Hobbies (25)
Home/Living (38)
Media (23)
Music/Movies (60)
Outdoors (16)
Regional (27)
Religion (17)
Sci/Tech (57)
Society (54)
Sports (36)
Crafts
Cars
Cigars
Dolls/Puppets
Guns
Magic/Illusions
Puzzles
Poker
Billiards
Board Games
Vintage Cars
Collecting
Gambling
Humor
Motorcycles
Satire
Memorabilia
Car Parts
Chess
Quizzes
Roleplaying Games
Knitting/Crochet
Photo Gear
Sewing/Quilting
Card Games
Save preferences

FIGURE 9-1. Manage your interests on StumbleUpon

are interested in, such as arts/history, health, outdoors, and religion. You can check as many categories as you want.

Now you're ready to start stumbling.

Using StumbleUpon: Toolbar basics

StumbleUpon is easy to use but may be more difficult to "master" if you are looking for community and trying to achieve social media marketing goals. Let's take a look at the toolbar and the various features available (Figure 9-2).

Stumble!
I like it!
Send to
Channels:
All
Favorites
Friends
Tools
1
2
3
4
5
6
7
8
9
10
11
12
13
14
15
16

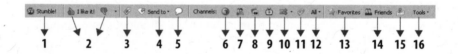

FIGURE 9-2. The StumbleUpon toolbar

The StumbleUpon toolbar includes a number of different icons, and even some of the heaviest users don't actually know what they all do. Here are some of the most useful and most commonly used icons.

Stumble!
: When you click the Stumble! button (1), the toolbar will spit out a brand-new web page that is relevant to your interests based on the information you have already provided to the service. In an effort to help you use the network for social interaction, StumbleUpon

places in the toolbar the username of any individual on the service (often someone in your network) who has also stumbled upon that story in the past (Figure 9-3).

You can click on this name and be directed to that user's StumbleUpon home page or blog, which will feature that user's latest discoveries and reviews.

FIGURE 9-3. StumbleUpon friend referral

Thumbs up/thumbs down

Do you like the page you were just presented with? If you do, thumb it up (2)! If you don't, feel free to vote against it with the thumbs-down button (2).

Tag

Here's your social bookmarking at work. When you click the Tag icon (3), you'll open a window that allows you to categorize your page with up to five unique tags (Figure 9-4). While StumbleUpon allows spaces in tags (delicious.com does not), you may prefer to be consistent with the rest of the users in your network and use spaces, dashes, or no spacing at all. If you are simply using StumbleUpon for your own personal bookmarking, there is no need to follow any consistent tagging pattern unless you do want to locate those sites again for your own purposes.

While tags do not seem to do much at this point, they are very interactive and can aid in content discovery.

FIGURE 9-4. StumbleUpon tagging window

Send To

The Send To menu (4) contains a list of users in your network. At first, it will be empty. You can begin building your network of users by subscribing to their favorites and adding them as friends. If they reciprocate, you can start sending your favorite pages to them. When you receive new incoming content via users in your network, a number will appear on your toolbar next to the Stumble! button (Figure 9-5). The best part is that when a

FIGURE 9-5. Your StumbleUpon friends have sent you two new stumbles

user first "stumbles," he will be forced to see your page and any other pages waiting in the queue for his exploration. He must see your pages before seeing other pages contributed by other StumbleUpon users. When you receive an incoming page, you will see a message at the top of the page with a short description sent to you by your friend. You can respond to or dismiss this message. While the Send To menu feature is quite useful, you shouldn't abuse the tool and send many pages frequently—don't overwhelm your friends with multiple Stumble requests. Additionally, try not to send too much content from the same URL to the same friends.

NOTE

Until recently, the only way you were able to share content with your friends was via the Send To button, and with that tool, you'd have to manually send content to each individual friend. StumbleUpon recently released a new Share feature that lets you send the same page to multiple contacts at once, thereby saving you time and energy.

Reviews of this page

The speech-bubble icon (5) points to StumbleUpon reviews of any specific page. When you click on this icon, you will find out if the page has already been discovered on StumbleUpon. You can either be the first one to discover it with your review, or if it has already been discovered, you can add your own personalized review. Normally, though, you should always "thumbs up" or "thumbs down" a page before (or after) adding a review to give it your endorsement (or disapproval), so that the StumbleUpon algorithm takes your opinion into account. A thumbs up helps ensure that more people see that content, whereas a thumbs down may stop a submission from spreading (if enough other people also give it a thumbs down).

TIP

One of the biggest mistakes committed on StumbleUpon is that users give glowing reviews to pages without actually thumbing up the content. The "thumbs up" is believed by the community to be the most important part of the StumbleUpon algorithm, so don't forget to include both a review and a thumbs up.

Channel buttons

The next few buttons refer to the types of content you receive on StumbleUpon. These are the different StumbleUpon Channels. Perhaps you want to see news articles or video exclusively. Or maybe you want to get a mix of content, such as photographs, news, and content relevant to a specific topic. The next few icons correspond to the various channels that you can tune into in order to see related content.

Stumble all

The icon that looks like a globe (6) lets you stumble all kinds of StumbleUpon content. This is the default navigational setting. When you click this icon, you'll be presented with video, pictures, news, and any other content that fits in with your interests.

Stumble favorites from your people subscriptions

The icon that displays two users (7) lets you start exploring pages that have been submitted or thumbed up by your friends. If you click this icon without first building up a network, StumbleUpon will suggest pages to you that were recommended by any member of the service. Once you start adding these users to your network (by subscribing to their favorites or by adding them to your friends list), this feature will provide you with links of pages they have discovered or endorsed on the service.

Stumble images

If images and photographs are your thing (and you don't want to read much), stumbling images (8) would be the ideal option for you.

Stumble videos

StumbleUpon also has a variety of video submissions that you can watch. Click the Video icon (9) to access StumbleUpon videos.

StumbleThru a website

Use this icon (10) to "stumble through" specific websites that have thousands of submitted pages. These websites are: BBC News, Blogspot, Break.com (a video-sharing site), CNN, Flickr, Funny or Die (a video-sharing site), HowStuffWorks (an informational site), Huffington Post (a political blog), MetaCafe (a video-sharing site), MySpace, PBS, PhysOrg (a news site), Rolling Stone, Scientific American, The Onion, university websites, U.S. government websites, Wikipedia, Wired, WordPress, and YouTube. There's no real benefit to stumbling within these pages, but you may prefer to discover specific types of content, and as such, StumbleUpon has made it easier for you to find user-recommended pages from a variety of popular sources.

Stumble news items

If you are interested in news, click the Newspaper icon (11) to be presented with only news stories. These are typically recent stories but not necessarily breaking news.

All ("Select a channel")

The All drop-down menu (12) gives you options to select any of these preceding channels (such as video, images, and friends' recommendations) and contains other features, including a Search tool. You can also only stumble content relevant to your interests (which you would have specified in the form shown in Figure 9-1). The menu is prepopulated with content that corresponds to your interests.

The next part of the toolbar contains account-specific features.

Favorites

The Favorites icon (13) takes you to your home page on StumbleUpon.

Friends

The Friends icon (14) shows you who your mutual friends are ("Your Friends"), as well as individuals whose pages you have subscribed to but who are not subscribed to you ("Your subscriptions"). On the sidebar on the right side of the Friends page, you will also see who is subscribed to your pages and find other users on the StumbleUpon service who are like you (Figure 9-6).

DEFINITION

On a social network, a *mutual friend* refers to someone who you have added as a friend who also has added you as a friend.

FIGURE 9-6. *Your StumbleUpon friends, subscribers, and similar stumblers*

Inbox

Click the Inbox icon (15) to access your StumbleUpon inbox. As you become more and more active on the service and broaden your network, you will surely begin receiving fan mail or requests to add other individuals to your network. You will also receive incoming messages when other users of the service review your profile page (and in most cases, they're doing so because they like you: you are providing very valuable stumbles!).

Tools

Click the Tools icon (16) to customize your toolbar, modify your account preferences, find more friends, log in or out of the service, and get more information about StumbleUpon.

Sharing your content with all your friends in one fell swoop

In the previous section, you learned that you can use the Send To menu to send a web page to a single friend. If you wanted to use the Send To menu to send the page to more than one person, however, you'd have to repeat this process again and again.

Fortunately, StumbleUpon addresses this problem with its Share feature, which lets you send a single page to multiple friends (all of them or a select few; see Figure 9-7) at once.

FIGURE 9-7. StumbleUpon's Share feature lets you send a page to multiple users with a personalized message

You can access this somewhat-hidden Share feature by going to your home page and hovering your mouse over a page you have already stumbled. Click the Share link in the bottom righthand corner. A list of all your mutual friends will be displayed. Select the individuals that you want to send the page to, and add a personalized message.

The benefit is that all your friends will definitely see your page the next time they log in to StumbleUpon. On the other hand, some users suspect that this behavior may minimize the impact of the StumbleUpon algorithm, especially when individuals give a thumbs up on every page they are alerted to. As such, it is possible that the pages may never leave the individuals in your network, even if they thumbs up the page, but this may be because friends often thumbs up every page they receive. If you are the recipient of a page from an individual in your network, don't give it the thumbs up unless you genuinely approve of it.

Stumbling and reviewing content

Now that you have an understanding of StumbleUpon's basic toolbar, you can start stumbling and reviewing content that you like. When you are the first person to give the thumbs up to a page, you will be presented with a StumbleUpon Discovery page (Figure 9-8), where you can add the first review for the submission. You can also categorize the submission by selecting one of the topics (or choosing a topic from the drop-down list). You can add related tags to this page and you can flag it as "Adult."

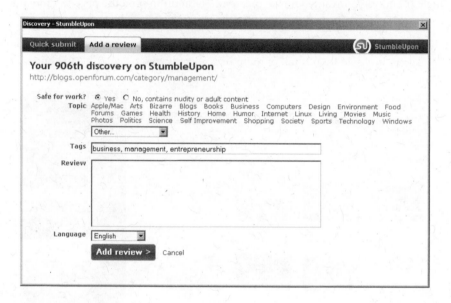

FIGURE 9-8. A StumbleUpon Discovery page

> **TIP**
>
> Typically, you want to avoid submitting "adult" content on StumbleUpon; by default, the service often categorizes adult-flagged content as pornography and will restrict most users from viewing these pages.

If you're aiming to build a credible StumbleUpon account, you should be active on the service. This translates to more friends and a stronger and wider network. As a solid StumbleUpon contributor, you should review about 10–15 pages a day and give the thumbs up on other content as well. You should also contribute good content to StumbleUpon and not all from a single source URL, since that can be perceived as excessive and overtly self-promotional. This activity will naturally cause other users to become aware of your profile and add you as a friend. It's also a good idea to look into the activities of other users, and to contact them directly or start subscribing to their submissions.

When you click the speech bubble on the StumbleUpon toolbar (see Figure 9-2), you'll be able to see reviews of your page. You can also see how many people thumbed up a particular piece of content (Figure 9-9).

FIGURE 9-9. A StumbleUpon review page shows the reviews (lefthand column) and the number of thumbs ups the page has received, as well as the avatars of the users who have endorsed the content (righthand column)

NOTE

In June of 2009, StumbleUpon introduced su.pr (*http://su.pr*), its URL shortening service. Like URL shorteners described in Chapter 6, su.pr works with Twitter, but it also integrates the finest features of StumbleUpon as well, showing submitters StumbleUpon reviews and clickthrough data with detailed graphs.

Understanding a StumbleUpon user page

A StumbleUpon user page (Figure 9-10) gives you information about a user and compares that individual to you in terms of the pages you like.

FIGURE 9-10. A StumbleUpon user page

At first, the StumbleUpon user page looks like eye candy: a grid (or list) of pages that the StumbleUpon user endorses.

1. The StumbleUpon user avatar is the picture the user uploaded to correspond with his profile. This is a great personalization tactic. Unlike other services, StumbleUpon lets you upload a large image as your avatar, so feel free to go crazy with this.

2. This section gives you more information about the user based on the information he provided to StumbleUpon. A user biography can be as detailed or as minimal as desired.

3. These graphics correspond to pages the user has reviewed. Some of these images are accompanied by stars indicating their popularity on StumbleUpon. The actual algorithm determining how a submission accrues stars is unknown.

4. You can find out more about the user (especially how active he is on the service) by reading StumbleUpon-specific information about him.

5. Based on the pages you like, how similar are you to this particular user? StumbleUpon uses a Venn diagram to illustrate similarity.

6. If you enjoy this user's submissions, maybe you would like to learn more about his friends and perhaps subscribe to their submissions as well. There's a list of those users on the righthand sidebar.

7. You can find out even more information about this user, such as when he last logged in, when he joined the service, and if he prefers video or photos.

8. This is a tag cloud that corresponds to the types of pages he likes most on the service. The larger the size of the tag, the more involved he is in reviewing and thumbing up pages with that specific tag.

If you are interested in taking the next step and establishing a relationship with a particular submitter, don't be shy. Feel free to contact him through his personalized page. The StumbleUpon community is a very friendly one, and the more active users are very responsive.

Interacting with tags

You've learned a bit about tagging, but how do you interact with them on StumbleUpon to make your life easier? First, it is helpful to familiarize yourself with the more popular tags (Figure 9-11). You can access these tags at *http://www.stumbleupon.com/tag/*.

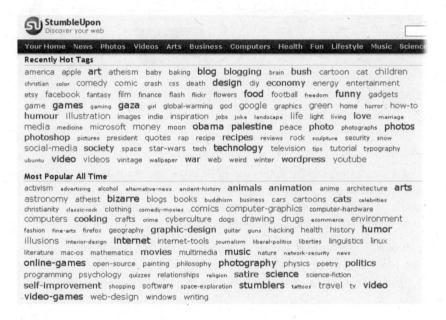

FIGURE 9-11. The StumbleUpon tag page

This page is divided into two sections. Recently Hot Tags are typically relevant to specific news occurrences (for example, the Mumbai terrorist attacks, the inauguration of Barack Obama, the Swine Flu, the Hudson River plane crash, and so on). The Most Popular All Time section shows the tags that are most popular, and will therefore work best on StumbleUpon. The page shown in Figure 9-11 indicates that the best types of content are usually funny, about cats, have eye candy (photography), and also cover bizarre news.

TIP

Image-intensive content typically performs very well on StumbleUpon. Since stumblers may not necessarily be reading the content, visuals help tremendously.

There are two very important ways to engage with tags on a consistent basis on StumbleUpon. The first is to use tags to discover new content. For example, refer back to the Tag page shown in Figure 9-11. For the purposes of this example, I clicked on "self-improvement" and was taken to a dedicated Self Improvement page (Figure 9-12). Next to that tag, a link displays "I like this." I can subscribe to this tag by clicking on that link, and then I will be presented with more content on that subject matter.

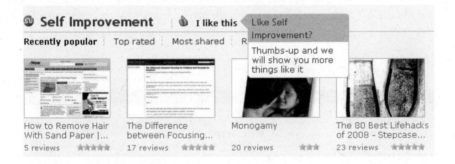

FIGURE 9-12. Subscribing to tags on StumbleUpon

You can also discover new tags by simply navigating to a page with a known tag, such as *http://www.stumbleupon.com/tag/food* or *http://www.stumbleupon.com/tag/crafts*. You can use this opportunity to network with individuals who enjoy content from those pages; the avatars of those users appear on the sidebar along the right side.

The best part of the tag feature is that you can find content that may inspire you to write your own articles, blog posts, or viral marketing pieces. Visit the pages that are linked from these tags and jot down ideas. You can also find relevant blogs for networking opportunities using this method.

delicious.com: One Social Bookmarking Site to Rule Them All

delicious.com is by far the most popular social bookmarking service. Founded in 2003 and acquired in 2005 by Yahoo!, today delicious.com boasts more than 150 million bookmarks. It has really taken the reins on social bookmarking.

While the service itself is mostly text-based, you can do a lot of social networking on it if you understand the service and how to use it. The best way to get started is to add the delicious.com Bookmarks add-on (*http://blog.delicious.com/blog/2008/05/internet-explorer-and-delicious*

.html), which is compatible with Internet Explorer and Mozilla Firefox. This adds three icons to your browser, as well as a new Bookmarks sidebar that displays your most frequently used tags (Figure 9-13). You can hide this sidebar if you wish.

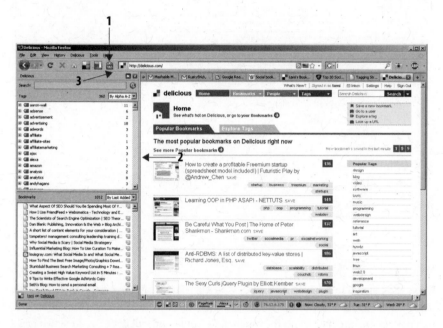

FIGURE 9-13. delicious.com icons and sidebar

To save a bookmark, simply click the Tag button on the toolbar and use the delicious.com pop-up window (Figure 9-14) to start your bookmarking. This window allows you to enter information about the URL.

As you can see in Figure 9-14, you can personalize your bookmarks. You may opt to use the provided title or enter your own (something more descriptive, perhaps). You can also add notes that will help you recall the page later when you are digging through your bookmark archives. Finally, you can tag the page.

Tagging is one of the most important parts of delicious.com, both for your own benefit and for the entire social network. delicious.com tags do not work well with commas.

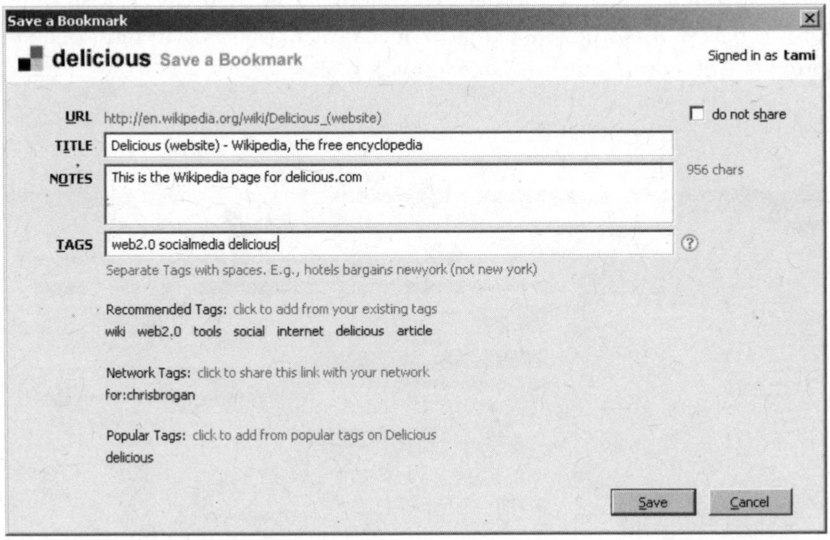

FIGURE 9-14. A delicious.com pop-up window for saving a bookmark

NOTE

If you use commas in a delicious.com tag, your tag will actually have the comma appended to it! This can get confusing, especially if you tag items with "marketing" (without the comma) and "marketing," (with the comma). delicious.com will not treat these two tags as identical; instead, they are treated distinctly. Further, spacing in tags is not a common practice; either treat the words as one (socialmedia) or use dashes to separate them (social-media). Even using quotes ("social media") will create two separate tags, one called "social" and the other called "media."

As you enter the name of a new tag, you will see a list of matching entries for tags that you have already created. For example, when you type "blog" in the tagging area, delicious.com will show you similar tags that you have already created, such as "blogging" and "blogs."

The tagging area also shows you another option for delicious.com: networking. The appropriate tag to share content with a friend is for:*username*. If you start saving content for another user, she will see it in her inbox (which is located at *http://delicious.com/inbox/username*). She can then choose to save the bookmark herself or to simply access the page.

Finally, you may want to utilize the "popular tags" field, which will give you an idea of how other delicious.com users are tagging the same page. In Figure 9-14, the popular bookmark is "delicious." This can give you an idea of how to tag the page, especially if you are using delicious.com for content discovery and trying to be consistent with the rest of the network.

You may also notice that you can choose whether or not to share your bookmarks. If you do not share them, they will be visible only to you. However, if you do share them, other users can view them your delicious.com home page (*http://delicious.com/yourusername*).

Navigating your bookmarks may become cumbersome once you start saving multiple tags for multiple pages. However, tagging will help you locate what you are looking for, so don't shy away from it; there's no limit to what you can tag on delicious.com. If you want to see your bookmarks at a glance, you can visit your tags page for a tag cloud (Figure 9-15).

FIGURE 9-15. A sample delicious.com tag cloud

As discussed earlier in this chapter, the sizing of the tags is determined by the frequency of times the tag was used. In the case of Figure 9-15, the boldfaced tags show that I save a lot of tags on social media (tag: socialmedia), marketing, blogging, WordPress, and SEO. I also have an interest in photography, Twitter, and plug-ins, as evidenced by the tags that are not bold but are larger than the plain-text tags.

This tag page is accessible to anyone. To view mine, visit *http://www.delicious.com/tags/tami*.

How delicious.com aids in understanding your brand

delicious.com works wonders for understanding how your thoughts are perceived. Anyone can navigate to my tag page. Anyone can navigate to my home page. You can also navigate to my individual tags and see exactly what I tag "photography." For example, am I looking at photographs and tagging them with "photography"? Am I searching for photography equipment and tagging them with "photography"? Or am I saving photography guides and how-to articles with the "photography" tag? Until you actually find out what I tagged as "photography," you will not know how I use that tag.

This leads us to an important consideration about tagging that is most relevant to delicious.com but that also comes up on other bookmarking services, such as StumbleUpon. How do people identify with your content? They may tag an article about your customer service as "excellent" or "customerservice" or "lessons." They may tag a bad publicity stunt as "pr" or "pitfallstoavoid." These tags can give you an indication of how people perceive an article and what they think about you. You can gather a lot of intelligence by looking at tags for an already-bookmarked story. On delicious.com, there is no end to the types of tags used by members of the service.

NOTE

The "toread" tag typically means that the person who bookmarked the article wants to read the content at a later time but was unable to do so when he bookmarked it. It's actually a heavily used tag on delicious.com. Beyond knowing that a person has an interest in the content, there's not much else to read into it.

How delicious.com aids in content discovery

Tagging can also give you content inspiration, which ultimately can help equip you with ideas for your own blogging and social media marketing. Further, this information can inform you of issues that are new and important in your particular industry. Timely news gets posted to delicious.com all the time, and thus, delicious.com is an excellent service to find new and interesting stories and articles that you may not discover via regular news outlets.

For example, if you navigate to *http://www.delicious.com/popular/socialmedia*, you will find a list of the most recently popular tags on social media. Often, you'll find strategy guides here and articles that may even illustrate missteps in social media marketing. You may also be linked to news articles that are relevant to the subject matter.

These articles themselves may not be as timely as you would like, but if you subscribe to the RSS feed for any particular tag, you're bound to get new content delivered directly to your RSS reader as soon as it becomes popular on the service. Thus, you can keep abreast of the most popular and latest social media developments—or progress in any area.

Certainly, you don't have to subscribe to only the popular tags. The same type of tag without the "popular" designation is available on delicious.com. If you want to see *all* content tagged "socialmedia" (popular or not), navigate to (or subscribe to) *http://www.delicious.com/tag/socialmedia*. You should keep in mind, however, that content is constantly being updated on the service, so you're likely to get bombarded with new updates, some of which may not even be relevant. You're bound to see foreign-language submissions and spam if you choose this subscription method, so perhaps it is better just to limit your subscriptions to the popular tags.

TIP

Want to find content that corresponds to more than one tag? Try *http://delicious.com/tag/tag1+tag2*. For example, if you want to find articles tagged both socialmedia *and* marketing, you can hop on over to *http://delicious.com/tag/socialmedia+marketing*. Unfortunately, this trick only works for individual tag pages and not for popular pages.

Social media marketing and delicious.com: What you need to know

The delicious.com home page actually sends a lot of traffic to the sites featured on the page, as do the individual popular pages. But how do you get there?

There are two main keys to success on delicious.com:

- Write good content that people will be compelled to bookmark.
- Have a strong enough network of friends using the service that they can bookmark the content for you, possibly by request if necessary.

To gain popularity on a particular tag page, you should encourage your readers to use specific tags (for example, plug-ins versus plugins or blogging versus blogs), though it is helpful to use the more popular tags on the service so that other active delicious.com users outside your network will also be exposed to your content, especially if they subscribe to specific tags. Again, there's really no right or wrong tag on delicious.com, but if you do want to use the service for networking and community, being consistent with the tags the rest of the community uses will be helpful.

It also helps to be the first individual to tag a page. Eventually, when the page reaches the delicious.com popular pages, the service acknowledges the person who posted the bookmark first. People may choose to network with you on the basis of your activity (the frequency of bookmarks posted to the service and how quickly you add bookmarks).

The delicious.com home page shows the most popular bookmarks on the site at any given time. This page is not updated dynamically; instead, there are updates to delicious.com every four hours. Depending on the time of day, you will need to have around 130 or more users bookmark your page in order to be featured on the delicious.com home page. The *http://www .delicious.com/popular* page does not require your site to have as many bookmarks to be featured, but you should get a lot of people to add your bookmark to the service quickly to be featured there.

Well-researched content typically performs best on delicious.com. Eye candy is not a requirement, nor does it have as strong an influence on making content buzzworthy like it does on StumbleUpon. The delicious.com community has broad interests; it's not necessarily seeking out the pretty pictures. Bookmarking on StumbleUpon and delicious.com differ in that StumbleUpon bookmarks are useful for the short term, but you're not likely to use the service for content archival. On the other hand, the content saved on delicious.com is typically something that you would want to reference in the future. The idea behind delicious.com bookmarking is consistent with that of traditional bookmarking: the majority of the users on the site like to save interesting and informational content. Other types of content are still saved, of course, though it depends on user preferences.

Diigo: A Promising Contender for the Researcher

Diigo (*http://www.diigo.com*) is a social bookmarking network. Users each maintain a profile page (Figure 9-16) somewhat similar in layout to Facebook, but the core idea behind the service is to aid in research and bookmarking. In fact, if you're already active on another social bookmarking network, Diigo may already have the tools to import your bookmarks; delicious.com, for example, is one of Diigo's supported sites.

FIGURE 9-16. The Diigo profile page

Diigo comes with an optional but very useful bookmark toolbar that lets you highlight and annotate portions of web pages. The best part about this feature is that when you visit the page again, these highlights and annotations will be still be visible. Diigo likens this service to physically highlighting a book. Additionally, the Diigo extension offers a sidebar that lets you browse your bookmarks, your friends' activity on the service, and more.

Diigo also lets you research specific domains for your own marketing goals. With Diigo, you can easily view statistics about your web pages. Which bookmarks are being saved under a specific site? Who is reading the content? What are their interests? Unlike other tools and social

FIGURE 9-17. Diigo provides site statistics, letting users know how many bookmarks and readers a site has

bookmarking services, you can view a single site's statistics easily at a glance (Figure 9-17), which is especially useful for gathering statistics and researching a particular domain name, such as your competitor's or your own.

Moreover, one of the most compelling reasons to use Diigo is that the tool searches the full text of entire web pages. This functionality is not offered by any of the aforementioned tools. If you are storing hundreds of articles covering a single topic on Diigo, you might want to carry out a full-text search within all of your bookmarks to find the very page you're seeking, and because Diigo stores the highlights, annotations, and text on its own servers, it delivers.

Diigo is not as popular as StumbleUpon or delicious.com, but it is well accepted as a research tool. Many active users of the service prefer Diigo over the competition because of its enhanced features.

Other Social Bookmarking Sites

StumbleUpon, delicious.com, and Diigo are the biggest social networks in the bookmarking sphere, but they're not the only bookmarking services available. If you want to expand your reach on multiple sites, you should leverage as many sites as possible. It's important to note that it will be difficult to establish yourself as a valued community member on all sites; it is better to be known as a valued contributor on one or two key sites instead of spreading yourself too thin. In fact, the other known social bookmarking sites do not require much involvement compared to delicious.com, StumbleUpon, and Diigo, but they are good tools to know about and understand.

Mento: A Small but Promising Tool

Mento (*http://www.mento.info*) is one of the newer social bookmarking tools. While probably the least known of all social bookmarking tools discussed in this chapter, it is a lightweight tool with all the features expected of social bookmarking services.

Mento works with the aid of a toolbar, though it's also possible to add bookmarks through the Mento web interface. The site also features a bookmarklet for even quicker bookmarking. Using the toolbar, you can send links to your friends, save to your personalized page (for example, *http://www.mento.info/yourname*; see Figure 9-18), or reply to an existing bookmark that one of your networked friends may have already saved. Mento emphasizes social networking more overtly than delicious.com.

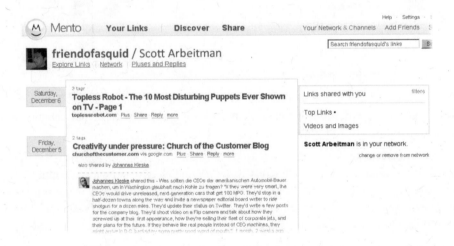

FIGURE 9-18. A Mento user profile

Twine: Interest-Based Information Discovery

Twine (*http://www.twine.com*) is another social bookmarking service that organizes information based on your interests. On the service, a "Twine" is a container that stores content (Figure 9-19). For example, you might want to store articles and videos about recipes to a Twine called "Food" or "Recipes." Each of these Twines has an individual email address, so you

FIGURE 9-19. A Twine container for writing

can send content directly via email to be stored in a Twine. Additionally, you can store content to a Twine via a bookmarklet.

Anyone can start a new Twine, and it can be private or public. The most popular Twines have the most content. Twines can contain files, videos, photos, web pages, and anything that has an associated URL.

If you choose to receive Twines via a daily digest, they will be delivered directly to your inbox. Twine allows you to discover new relevant content in a timely fashion, though subscribing to the content via RSS is also available. Many users believe that Twine brings a lot of new and interesting content to the table that is not easy to find on other social networks.

Mister Wong: Multilingual Social Bookmarking

Another social media site is Mister Wong (*http://www.mister-wong.com*) (see Figure 9-20), which lets users save bookmarks with relevant tags. When you view a bookmark, you will discover other bookmarks that are saved within the same domain (up to 30), so you can discover other popular content originating from that URL.

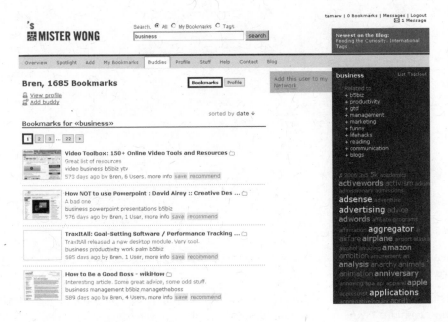

FIGURE 9-20. A user page on Mister Wong

Mister Wong's most common categories (as evidenced by the biggest tags) are business, marketing, free stuff ("for free"), home, Internet, online, and video. Crafting content for these categories will likely boost your popularity.

Additionally, Mister Wong awards its biggest supporters. On the sidebar of any specific tag page, you can view a list of "Power Users." These users have saved many bookmarks with the same tag. Active users on the service can be featured as power users.

Summary

Social bookmarking has evolved from a private to a public phenomenon. In the future, we may see an even more widespread evolution as search engines start using personalization and tagging to find the best results for any particular searcher.

Today, with the aid of numerous online services, bookmarking is a very social activity. StumbleUpon and delicious.com are the two biggest tools for this purpose, but Diigo is an up-and-coming service that offers all of the benefits of the popular delicious.com service and more.

StumbleUpon is really a social site at its core. Based on your interests, you will be presented with pages that other users with similar interests have already recommended. You can also begin building up a network of users of the service. If you share a page with an active user and she in turn promotes the same page, you have the potential to give a single page thousands of page views and plenty of exposure.

delicious.com, on the other hand, really does focus on bookmarking at its core. Tagging is a big part of the success of the service, and you can navigate to tags to find relevant content. There's a social networking element to delicious.com, but it's not as prevalent and is not used as much as StumbleUpon.

There are numerous other small social bookmarking sites that are not as widely known but can be useful for social media marketing and for personal bookmarking. They include Mento, Twine, and Mister Wong.

Endnotes

1. *http://www.searchengineland.com/search-40-putting-humans-back-in-search-14086*
2. *http://www.googleblog.blogspot.com/2007/02/personally-speaking.html*
3. *http://www.googleblog.blogspot.com/2008/11/searchwiki-make-search-your-own.html*
4. *http://www.techcrunch.com/2008/12/10/google-search-wiki-to-soon-include-an-off-button-thank-you-marissa*

CHAPTER TEN

Social News Brings You Page Views

Is all news interesting? Chances are, you don't read your newspaper cover to cover, so the answer is no. On the other hand, you might be deeply affected when you read a news article sent to you by a significant other, colleague, or friend; if it's being shared, the chance of it containing something exciting is likely higher than the same old news stories. Social news has evolved on the premise that individuals flock to the attention-grabbing stories and headlines. The most appealing new stories should have greater exposure. People like sharing articles that are humorous, bizarre, or informative with their friends, so why not create a website where the most interesting news stories are easy to access?

Social news sites are democratic websites that allow individuals to vote on interesting content that they want the rest of the sites' user base to see. Exposure on social news sites can bring hundreds to thousands of page views, which may not only create awareness of your product, but may also lead to significant company change. If nothing else, however, these popular social news stories evoke strong emotions, make people aware of who you are, and cause people to talk about you.

The Wisdom of Crowds

Many of us probably believe that we're smart, but that only a few select people are truly brilliant. We may be lucky when we guess the right answer to an impossible question, but there are a handful of individuals who are true geniuses.

Or so that's what you may feel compelled to believe.

James Surowiecki's 2004 masterpiece *The Wisdom of Crowds* (Anchor) discusses and disputes this hypothesis. The book introduces British scientist Francis Galton, who conducted an experiment at a weight competition held at the 1906 Plymouth County Fair. Participants at the event were asked to place wagers on the weight of an ox after it was slaughtered and dressed. Nearly 800 contestants placed their bets on the ox. Many of the contestants were experts and presumably possessed the intelligence required to accurately assess the animal's weight. A large number of common folk participated in the competition as well.

In this experiment, Galton aimed to prove that the experts had the most accurate knowledge of the ox's weight. He collected each individual's guess and performed a detailed statistical analysis. After amassing the data and performing his calculations, Galton was surprised to find out that not a single expert precisely guessed the ox's weight of 1,198 pounds. However, when all 800 guesses were averaged, the result was an astonishingly close 1,197 pounds.

After the experiment, Galton concluded that democratic judgment was trustworthy. The thesis of Surowiecki's book is that "under the right circumstances, groups are remarkably intelligent, and are often smarter than the smartest people in them. Groups do not need to be dominated by exceptionally intelligent people in order to be smart. Even if most of the people within a group are not especially well-informed or rational, it can still reach a collectively wise decision."

Social news operates on a similar premise: the Web has been democratized. Forerunner social news site Digg explains the democratization of the Web succinctly (*http://digg.com/about*):

> You won't find editors at Digg—we're here to provide a place where people can collectively determine the value of content and we're changing the way people consume information online.
>
> How do we do this? Everything on Digg—from news to videos to images—is submitted by our community (that would be you). Once something is submitted, other people see it and Digg what they like best. If your submission rocks and receives enough Diggs, it is promoted to the front page for the millions of our visitors to see.

Clearly, if the crowds determine that content to be worthy, it will be promoted to the front page of the site to receive thousands to millions of visits.

Social News Is "Social"

While the phenomenon of "wisdom of crowds" sounds promising, the idea of social news is that it is entirely social. Galton's experiment succeeded because 800 different individuals offered their honest insights into the weight of an ox without consulting with one another. On the other hand, social news sites, by default, consist of individuals uniting under common interests. More importantly, the users of these social news sites are actually members of a community. Success in social news rarely happens by chance; success comes from understanding expectations of these community members and following proper networking procedure.

Once upon a time, it may have been the case that a newcomer on a site like Digg could rise to the top simply by letting "other people see [the content] and Digg what they like best." Before communities actually materialized, democratization of the Web made all the sense in the world. But with evolution of social news sites comes change. Herd mentality becomes increasingly dominant. This was especially apparent during the Digg revolt of April 2007, when a user on the site posted a highly controversial and proprietary HD-DVD processing key that was illegally obtained by a computer cracker. Giving in to pressure from legal counsel, Digg decided to remove all submissions referring to the sensitive data and CEO Jay Adelson explained that the social news site is required to abide by laws, especially as they relate to infringement of copyrighted property.[1]

The users weren't satisfied. The Digg staff worked overtime to remove the HD-DVD processing key, but brand-new and related submissions were being added to the site every minute. All in all, thousands of blogs, articles, and forums referenced the key and attributed it to Digg. Eventually, Digg's upper management realized that it couldn't fight against its constituents— too many users were posting the illegal code to the site despite the company's good-faith efforts to take the content down—and surrendered to the masses in a blog post, in which founder Kevin Rose said that the company would no longer bow down to the larger pressures and would instead listen to its users:[2]

> But now, after seeing hundreds of stories and reading thousands of comments, you've made it clear. You'd rather see Digg go down fighting than bow down to a bigger company. We hear you, and effective immediately we won't delete stories or comments containing the code and will deal with whatever the consequences might be.
>
> If we lose, then what the hell, at least we died trying.

This scenario demonstrated that social news is incredibly powerful and that users want to be heard. Community members had to work together for the revolt to be as successful as it was; they drew strength from one another and showed how they could essentially overtake a site that still seemed to possess authority.

The Digg revolt of 2007 (Figure 10-1), probably the biggest turning point for social news as a whole, shows that people will work together to share content with their peers, and perhaps most importantly, they will enlist those peers to help increase visibility for that content. Whereas social news sites are bound by laws and algorithmic restrictions, it is possible to prevail on these social news sites as long as you put a concentrated and dedicated effort into understanding the site's goals, networking with the proper individuals, and crafting the right kind of content that site members would like to read.

It may be true that individual users possess editorial power to influence the content that will be visible on the site, but it is becoming increasingly apparent that as a community, users can more effectively manipulate the visibility of articles on social news sites, and as such, the process is no longer entirely democratic. While every single participant has an equal voice on social news sites, if users band together to achieve or deny visibility, they will inevitably prevail.

FIGURE 10-1. The HD-DVD Revolt on Digg, May 2007

What Is Social News?

In the beginning, social news may have been a news article that you wanted to share with your friends and family. That bizarre story about a gorilla giving birth on a plane while in transit to a new zoo may have evoked in you emotions of surprise and glee, and you want to share the story with your friends who love animals. On the other hand, the unfortunate story about a toy that is killing animals but has not yet been recalled may evoke anger and resentment; your disgust with the manufacturer of that toy may prompt you to share that story with as many people as possible so that perhaps someone in your social circle can appeal to the manufacturer to reevaluate its products and consider pulling the toy off shelves. After all, sharing these stories means that an even wider audience becomes aware of the news. In social news, interesting content deserves to be shared.

But social news is not necessarily all about news stories any longer. Today, social news sites dedicated to highlighting these peculiar news events and informational stories have begun amassing other types of content, from hilarious videos to amusing pictures to detailed how-to guides on any subject. Articles can originate from respected and reputable sources such as the *New York Times* and the *Wall Street Journal,* or they can be submitted from even the smallest and newest blogs. Videos and pictures can come from any source on the Internet, from a video-sharing site (like YouTube) to a web page. There is typically no restriction on the type of source for the content on a social news site. By and large, social news sites like Digg (Figure 10-2) highlight the cream of the crop in web content. Further, just about any content can go on a social news site, as long as it is compelling and remarkable.

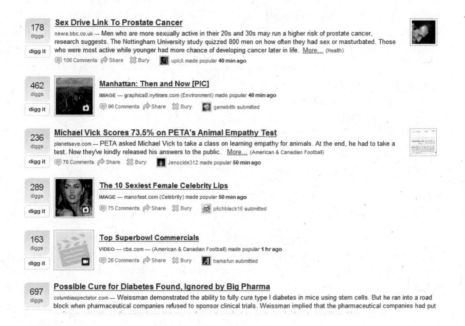

FIGURE 10-2. Digg's front page shows a unique mix of compelling video, pictures, and news

If you've ever visited a social news site, you are likely to stumble upon information that is unique. You are likely to visit these sites again and again to find more exceptional content that you would feel compelled to share with others. The titles and optional descriptions can compel you to read more about the story, and when you do read the article, you may want to save it or send it on to others. These stories can capture an audience's attention, help you reach hundreds of thousands of page views, and build hundreds of links that point to the original article.

What Are Social News Websites?

At this point, you have an understanding of social news and a brief overview of social news sites, but how exactly do they work? Social news websites, like social bookmarking sites, are communities that let you submit web content in the form of articles, videos, or pictures. In contrast to social bookmarking sites, however, where you may choose to bookmark your favorite website (which may not be a standalone article), the best-performing content on social news sites (and the content that is recommended) is interesting and appealing to a wide variety of users. In other words, it's not socially acceptable on social news sites to submit a whole blog or website, whereas that behavior is appropriate on social bookmarking sites. Usually, social news sites offer standalone articles laden with information and detail.

In a social news website, content is submitted, often reviewed by the system for duplicate entries, and then approved or rejected by the system. When a submission is accepted, a voting system comes into play. The submission is visible to the social news site's community at large, and members can either vote for or against a story or comment on it. After a certain number of votes have been accumulated, the story may be promoted to the front page of the site to be exposed to a much wider group of individuals, including those who may not venture to the inner pages of the site.

DEFINITION

Stories that are promoted to the front page of social news sites are considered "popular" in social news jargon.

Promotion to the front page is the ultimate goal of using social news sites for a variety of reasons:

- Ardent community members ("power users") of these social news sites take pride in this type of exposure, as it boosts their account profiles and establishes them as solid contributors to the site, especially if their submissions reach the front page repeatedly.

- A website that reaches the front page of a social news site gets traffic (at the minimum, it will get thousands of page views in a short period of time), and if the content is really interesting, visitors who stumble upon that submission may choose to write about and link to the story. Therefore, there is substantial opportunity to gain additional completely organic links through this method.

- Stories promoted to the front page of a social news site often are accessible from other websites. In addition to the links these stories will garner from bloggers and other writers, they will inevitably be submitted to other social networks and social bookmarking sites as other community members stumble upon the content and wish to share it.

There are hundreds of existing social news sites of all different types and sizes. Some cover all angles of news, from business to technology to funny pictures found on the Internet. Smaller niche-specific sites revolve around topical submissions, such as women's issues or financial news.

The biggest social news websites receive hundreds of submissions on a consistent basis. Therefore, your potential for high visibility may be slim on these sites, especially without considering all of the factors that can determine success on a social news site.

Factors That Influence Social News Front Page Promotion

To combat inflated voting and to address the power of the herd, social news sites operate with simple or complex algorithms that define the popularity of each story. The algorithm accounts for distinct votes but may include other factors, such as the number of negative votes made against the submission. The bigger and more popular social news sites that boast millions of unique visitors per month are required to make major algorithmic adjustments on a semiregular basis. On the other hand, smaller niche sites are typically built on open source platforms, such as Pligg, and cannot compute complicated algorithmic equations, so typically stories that reach a certain vote threshold (for example, 25 positive votes) will automatically be promoted to the front page. Once these sites are configured by their respective niche site owners, they may be customized and programmed with other safeguards, such as the ability to remove stories that are spammed or to negate a vote on a story with a "bury" vote.

Front page promotion on social news sites is typically dependent upon numerous factors. These considerations, described next, will vary depending on the site.

The user

On more popular social news sites, the username of the submitter is an important concern. Is the user known? Does he have many friends in his social network? Is he a new user, or is he a well-established community participant? Does he submit many stories on a regular basis, or do submissions come few and far between? Depending on how active a user has been on the service, subsequent success may be a lot harder to achieve. In other words, if you submit five stories to a social news site and they have all been promoted to its front page, your next few submissions may not be as lucky. Sometimes, you may find it necessary to take a break and focus your submissions on periods of time when you may be able to monitor your progress more closely. At the same time, your repeated success may require you to start getting more votes from more unique users on the service. Your first submission may have hit the front page with a mere 70 votes, but you may need 200 or more in the future as your name becomes known.

Does this mean that if you're an unknown user, you cannot achieve success on social news sites? A large percentage of active users will simply follow the more popular members, but social news sites try to ensure a level playing field for all participants, and therefore, it's anybody's game. However, it is generally agreed that an unknown user will need to work harder to achieve front page fame.

The diversity of voters

Many popular social news sites come equipped with algorithms that favor diversity. As you become more active on these social news sites, you'll see that the same few people may follow your submissions and vote on your stories. You may also be contacted directly by members who want to expand their networks by adding you as a friend. The bigger social news sites already know that people are networking onsite and offsite to possibly "game" the system. If the same few people vote for your stories all the time, their votes won't be valued as much by the algorithm as those from other community members who typically do not vote on your content.

The number of votes

The number of votes your story collects over a period of time compared to the average number of votes on other stories may determine how and if the story will reach the front page. However, the other factors discussed in this section may cause a delay in front page promotion. For example, your story with 235 votes on the top of Digg's upcoming queue is probably higher than the number of votes on average, but the particular submission may have other penalties counting against it, which is why that story is not yet on Digg's front page.

Speed of votes

The speed of votes is an important consideration for front page promotion. On Digg, for example, normal stories gain an average of 10–20 votes an hour, which is expected of stories on this social news site. However, breaking news stories will accrue votes extremely rapidly; for example, when actor Heath Ledger passed away at age 28, the story was promoted to the front page of Digg within hours because it received hundreds of votes in a very short time. In fact, the story submission had accumulated thousands of votes by day's end and now has over 23,000 unique Diggs. On the other hand, slower-moving stories sometimes do not see front page visibility at all. If your story has only garnered 140 votes in a 24-hour period, it may not ever be promoted to the front page. On the other hand, if your submission garners 140 organic votes in two hours, there is a strong probability that the story will be made "popular" and moved to the front page.

On smaller niche social news sites, if your story is promoted to the front page very quickly, gaming is suspected.

The story's category

On mainstream social sites, if you are submitting your story into a high-traffic category, it may be harder to achieve front page fame. Consider submitting stories to categories that are still relevant but do not have as much traffic, as there is less competition and not as many unique votes are required. On the other hand, on some social sites, the categories with less traffic may not be seen at all, so it is in your best interest to understand the community and the content it likes so that you can determine the best category for your submission.

The number of negative votes

Most, if not all, social sites have a way to give a negative vote to a submission. If your story collects too many negative votes, the story may never become popular and could be removed from the system entirely.

Unfortunately, not all social voting sites will disclose the number of negative votes. Digg, for example, will tell you how many positive votes a story has received, but the number of buries (negative votes) remains a mystery. There are third-party tools to find out the exact number of buries on a particular Digg submission, though there is debate about whether these tools are accurate.

The number of comments received

Each social news site submission contains a text area that allows users to comment on the story. Some sites even allow you to vote for or against these comments. In general, more comments on the story can help push it over the edge. However, users of social news sites should exercise caution and offer comments of value. Merely saying, "Thanks for the great article," will often cause other users to bury or give a negative vote to the article.

The time the story was submitted

If you're expecting your social news submission from three weeks ago to hit the front page, think again. Most social news sites have an algorithm that places emphasis on when the story was submitted, since the idea behind the "news" in "social news" is that it should actually be timely and recent. You'll often see stories become popular within 24 hours of submission, though occasionally you may also see stories from that were posted within 48 hours become popular. After that, your chances for front page promotion are much slimmer. It's not worth it to invest time and energy trying to get votes on stories that are past their prime. In fact, if you put a social media profile badge on your website (for example, a "Digg this!" button), you may want to remove it if the story didn't gain initial traction. It doesn't always look good to have a story from weeks ago that has only 5–10 votes. It certainly doesn't look good for the story to have a "0 Diggs" badge on the site, either, so be sure not to plaster all your blog posts or site articles with the Digg button. Only add it manually.

THE COST OF FREE TRAFFIC

If you've never experienced a surge of traffic on your website from a popular news site and are expecting that your social media marketing piece will perform well, be forewarned: your hosting provider may not be able to accommodate the traffic rush. Be sure to speak with your hosting provider to make sure it has appropriate safeguards in place and can anticipate that your website may be working overtime; the first one to three hours can show thousands of page views—perhaps hundreds per second, which is not very normal for a nonenterprise hosting environment (and may not even be expected of an enterprise infrastructure, either!). The biggest gripe of many social news users is that the "Digg effect" or the "Slashdot effect" has taken websites down and even contributed to terminated hosting contracts. Don't be one of the casualties. You may have invested hours in that social media promotion piece, but if your hosting setup can't handle it, your efforts will be fruitless.

Winning on Social News Sites

Now that you are armed with relevant information on what works on social news sites and what doesn't, how do you promote that content to a social news site? How do you use a social news site to achieve your social media marketing goals?

Build your power account

Recall that in Chapter 3, we discussed the 10 commandments of power account submitters. They are:

1. Thou shalt distinguish thyself with an avatar (Figure 10-3).
2. Thou shalt be genuine.
3. Thou shalt network.
4. Thou shalt submit high-quality stories to the social sites.
5. Thou shalt be fast.
6. Thou shalt study the sources that have achieved greatness on social sites to understand what the community likes.
7. Thou shalt dedicate time to the task.
8. Thou shalt help thy friends.
9. Thou shalt use consistent account names over all social networks.
10. Thou shalt use other social networks for inspiration.

These 10 commandments are a starting point for how you should conduct yourself on social news sites. It's important to be noticed with an avatar and to give to back to the community with genuine contributions. As you will see, if you engage yourself on a social network but do not network, you will find that success is hard to come by. After all, if you do not actively make people aware of your submission, how are other members supposed to find your specific social news story in a sea of hundreds of other similar stories?

More than that, know what the community wants. High-quality stories are a must. Study the front page of social news sites to find out exactly which websites by themselves are successful. Certain blogs are popular on certain social news sites, for example; if you want to grow your social news power account, you should know exactly which sites are generally popular on these sites and closely follow new updates to these blogs. If you're the first person to submit a powerful piece to a social news site, you're almost guaranteed to have a front page story under

FIGURE 10-3. *My Mixx profile sports an easy-to-remember avatar*

your wing. If you do this consistently, your power account is all the more powerful, and community members will take note. Be advised, though, that there are numerous users trying to be the first to submit a story to a social news site. Therefore, competition is fierce. As such, you may not be the first one to submit the story, but it doesn't hurt to try. In fact, if you consistently do beat all odds and are the first person to submit that story more than once, you will find that you gain followers by virtue of being a community participant who submits valuable content.

You will begin to see that networking is critical on these social news sites. While algorithms will prevent the obvious gaming, the more diverse your friends are, the more opportunity you will have to see your story hit the front page. The same few people should not always vote for your stories, as it then becomes obvious that they are helping you achieve your goal (and these users may be viewed unfavorably by other members). Voting rings and users who work toward front page visibility are often penalized, either by termination or by having their votes discounted (though not visibly) on these story submissions.

However, if you're an active participant, and you should be, you should vote on other stories, too. It won't be in your best interest to only promote stories that your friends are submitting. Look at other submissions in the sites' upcoming queues (Figure 10-4) and vote on the stories that are most interesting to you.

TIP

The upcoming queue is a great way to seek out new contacts to add to your growing network of friends. If you want to be noticed by top submitters, vote on their stories before they get popular. These users will take notice and will likely reciprocate, thereby boosting your account profile as well.

Navigate to individual category pages and vote on other stories that you feel deserve more eyeballs. Never focus your energy on only your friends; keep your voting behavior as natural as possible.

FIGURE 10-4. The upcoming queue on Internet marketing social news site sphinn.com

Comment on submissions as well, especially if you have something valuable to add (though snarky comments and humorous ones prevail on some sites). Comments help you establish your own credibility as a submitter, and depending on the site, you can gain respect among site members through the types of comments you provide. Further, submissions with valuable comments typically help bring the story to the front page.

Networking is something that needs to be emphasized heavily on social news sites. Many social news sites will let you fill out a profile to volunteer personal information so that people can network with you offsite. This profile may only allow you to write a short two- to three-sentence bio, but other sites allow you to fill out your Instant Messenger (IM) information or website URL so that people can learn more about you and contact you outside of the social community (Figure 10-5). Ultimately, though, this can work to your advantage: if you're in desperate need of a vote, you can almost always count on your friends for a helping hand!

FIGURE 10-5. Finance social news site tipd.com's profile lets you provide your URL and instant messaging contact information

There's a catch to this kind of behavior, however. It's important never to network with someone out of the blue and just ask her for a vote. Build a relationship first and foremost, and let the vote requests come only after you show a genuine interest in the individual you're soliciting votes from. Additionally, if you volunteer your personal information, do not be surprised if you get bombarded with requests. I have a fair number of friends who no longer use IM on a regular basis because as soon as they go online, they get hammered with requests to vote on stories. Instead, don't volunteer your IM name at all, or use a social media IM handle that you only give to your social news friends.

The bottom line is that success on social news sites takes time and dedication. You will need to allocate time to study successful stories on social sites and understand intricate details of what works and what doesn't. You will need to build up your network, and this will not happen overnight. Many of the active social news site participants are skeptical of new "friends," so you'll need to prove your worthiness first. You will need to spend a substantial amount of time at first to vote up content and to show that you mean business. Depending on the type of site, this could range from about an hour to several hours a day. However, if you are serious about building a power account, there is substantial ROI to be had by establishing yourself as a valuable community member, from being contacted by outside parties to submit content (for free or for payment) to building credibility and respect within these social networks. As an added bonus, your submitted content will have a much higher chance of front page success.

What if you can't build your power account? If you simply do not have time to spend on building a solid power account, find someone who has already established himself as a respected community member. Reach out to this person and ask him if he will submit the content on your behalf. This content will still need to adhere to the rules of acceptable social news content; top submitters value their reputations and won't submit just any content to these sites. In the next section, we'll discuss what types of content perform well on social news sites so you can get a good idea of what exactly to present them with.

Build the right content

Crafting the right content and having the right kind of submissions is vital for success on social news sites. From concept to completion, every single element of social news should be carefully reviewed. In general, the best-performing content pieces will have one or many of these elements:

Lists

Social news communities love lists (Figure 10-6). They're short, easily digestible, and are often informational enough that they can be referenced again and again (and therefore, they work very well on social bookmarking sites, too!). There are really no rules for lists, though you should dress them up with visuals and graphics. Typically, the best lists are meaty enough that there's an introductory sentence (in bold) followed by a more detailed explanation. One-liners are not perceived to be as well researched.

FIGURE 10-6. A popular list on women's social site kirtsy.com received 97 votes

Breaking news

It's a given that people want to find breaking news on social news sites. Whether it is an announcement of a plane crash impacting celebrity figures, the launch of a brand-new, highly anticipated green product, or news of an upcoming presidential election, social news still accounts for regular news, and these stories are very popular on social news sites. If you can be the first person to break such news, all the more power to you. If you can't break the news, perhaps you can be fast enough to craft interesting content in response to the breaking news (or in appropriate circumstances, to turn it into a humorous parody). See Figure 10-7 for an example of breaking news.

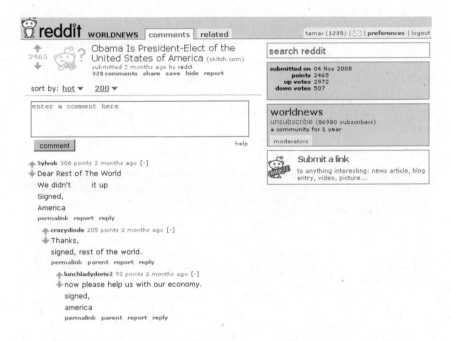

FIGURE 10-7. Obama's presidential win receives thousands of votes on reddit

Games

Can you create an addictive and inventive game that will cause users to come back for more and more? While social news sites do not have many of these types of submissions, probably due to high overhead costs of development, many viral games perform well on social news sites (Figure 10-8), especially if they're memorable. Quizzes also perform rather well.

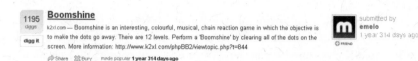

1195
diggs

Boomshine

digg it

k2xl.com — Boomshine is an interesting, colourful, musical, chain reaction game in which the objective is to make the dots go away. There are 12 levels. Perform a 'Boomshine' by clearing all of the dots on the screen. More information: http://www.k2xl.com/phpBB2/viewtopic.php?t=844

Share   Bury   made popular **1 year 314 days ago**

submitted by
emelo
1 year 314 days ago
FRIEND

FIGURE 10-8. Over 1,100 diggs have been collected for this game submission

Controversy

News that induces strong emotions often performs well on social news sites. The controversy may be about new restrictions imposed by an Internet company on bandwidth and privacy (such as net neutrality, a highly contested principle on how electronic communications should remain free of restrictions and censorship), or a story about a large corporation slighting an innocent and unsuspecting customer and causing her much harm and embarrassment. Controversy can also be more large-scale and related to national news in the political arena. If it's relevant to the community and the story is controversial in nature, it is fair game (Figure 10-9).

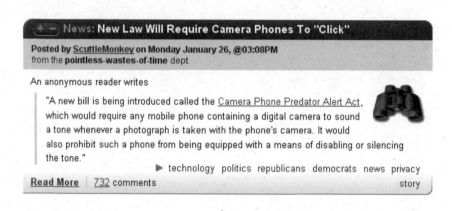

News: New Law Will Require Camera Phones To "Click"

Posted by **ScuttleMonkey** on Monday January 26, @03:08PM
from the **pointless-wastes-of-time** dept.

An anonymous reader writes

"A new bill is being introduced called the Camera Phone Predator Alert Act, which would require any mobile phone containing a digital camera to sound a tone whenever a photograph is taken with the phone's camera. It would also prohibit such a phone from being equipped with a means of disabling or silencing the tone."

▶ technology politics republicans democrats news privacy

Read More | 732 comments   story

FIGURE 10-9. A new law proposal caused slashdot users to go up in arms; over 700 site members have had their say

Videos

Have you ever come across a silly video or one that causes you to gasp in horror? Has a candid video of someone falling off a ladder and landing smack on his face caused you to cringe? What about a zany video with a cat singing a song? These aren't unrealistic expectations of video now that we're all content producers, and the craziest videos are often very successful on social news sites (Figure 10-10).

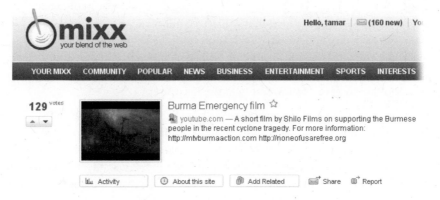

FIGURE 10-10. A popular video on Mixx has over 100 votes

Pictures

Like videos, funny pictures or "pictures that speak a thousand words" perform optimally on social news sites (Figure 10-11). On some social news sites, pictures can be labeled (PIC) in the title. On others, you can attach a picture to the submission. It is helpful to make your submission title clear, so make sure you indicate that the submission is a photograph or image.

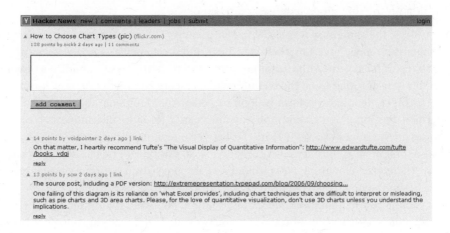

FIGURE 10-11. A picture submission on social news site Hacker News (YCombinator)

These are just a few types of content that typically perform well on social news sites. Since the type of well-received content is forever changing, study the front page of the social news site you intend to target to determine whether you will be successful in your content creation method. Most social news sites have an RSS feed of the popular stories, and this type of research does not require you to be active on the actual site at all times.

A strong title and description (where applicable) is crucial. Describe exactly what your submission is about. Write something that will drive your reader to click on the story for more information. Remove the name of the publication from the submission title and rework the title to be unique and more interesting than the original. You may find a great submission titled "Ways to Beat the Recession," but you could make that title more interesting by rewording it to "23 Unique Ways to Save Money and Beat the Recession."

Strong titles and descriptions also prevent stories from being flagged as spam. If you opt to use the title that the social news site suggests for you, the entire title, if not correctly rewritten (for example, "Ways to Beat the Recession >> The Money Blog on Finance, Money, and Finance") may look like you're submitting spam to the social news site. Put some effort into your submission. Make the community members of the site aware that your submission is worthy of being viewed, and that it is actually relevant to the site and the category to which you submitted it.

The domain on which you submit your content is important. You've learned that blogs and mainstream news sources are great places to find good content. How about your own marketing goals? Whatever the case may be, never use your sales site for social media marketing. If your page header displays "call us at 1-800-BUY-THIS," your motivations for social news promotion will be clear. Some social news sites will remove these submissions and ban the URL from being submitted to the service in the future. Instead, modify your page template and remove the overt commercialism from these web pages to be more social news-friendly, or push the content on a different domain or site.

In Figure 10-12, the content is very social media-friendly. The problem, though, is that this type of submission spells "overt marketing tactics," and social media-savvy users may be inclined to report it as spam. Although it performs well by itself, there should be a disconnect between this type of content and the website where you are aiming to gain new business and customers. This content may perform well on social bookmarking sites and even on StumbleUpon, but social news community members are a lot more discriminating, especially since stories on the front page are few and far between, and many people are viewing and judging the content all at once (whereas on other sites, this type of rapid visibility is not as common). Therefore, scrap the 1-800 number and any navigation information that indicates that there is some payment plan required. In the future, after the story's popularity has waned, you may want to restore that promotional information, but avoid submitting the story with that information readily visible—social news users will likely not want to endorse the content with their votes if a marketing goal is clearly the reason for your submission.

FIGURE 10-12. *Do not submit stories like this to social news sites!*

NOTE

In June 2009, Digg announced the imminent launch of Digg Ads, which allows advertisers to finally participate in Digg without the need to find a power submitter or become one. The concept behind them is that advertisers can promote advertisements that users can vote on. The submissions are marked as "Sponsored" listings. The more the ad is voted on, the less an advertiser has to pay. With a sponsored listing, that 1-800-BUY-THIS header may not have to be removed, but you should keep this advice in mind as it could result in less clicks (and more payouts to Digg).

When you study the popular submissions on any specific site, and in particular, the social sites that are almost impossible to market on, you'll have an idea of exactly what kind of packaging is required of your social news submission. Sometimes font size and presence of images is critical. Sometimes the placement of your social media promotional badge may need to be moved. Take a strong look at the content that has been promoted to social sites and make sure that your web development team accounts for considerations in site design. Similarly, look at the content to understand exactly what kind of subjects are preferred by the majority of site members. You'll soon notice that, depending on the social news community, certain news outlets are strongly disliked (FOX News, for example, is normally not a trusted news source by social news participants) and some topics are very popular (for example, Google and Apple stories are widely accepted on social news sites, but obvious "how to market your product" or "how to optimize your website for search engines" would likely never be acceptable). These trends can change over time, so be sure to review the latest popular submissions to see how the crowds react.

Can social news promotion really further my social media marketing goals?

Absolutely! Promotion within social news channels not only yields page views, but can also translate into powerful conversions. This may not happen solely by virtue of the story hitting the front page of a social news site, as not all social news users are looking to buy a particular product (though the right product may find new owners). The front page promotion is just the tip of the iceberg. It's the domino effect that causes success in social news sites. When social news users share their content with other audiences, the potential for conversions is much higher. In one case, after a particular story was promoted to a social news site from a travel company's website, the story saw over 50,000 page views and garnered over 1,000 links. Ultimately, bookings with that travel company were attributed directly to that social media marketing piece. The bottom line was that once the company was promoted in the social media marketing sphere, the content was on someone else's radar. Once the initial visibility is there on a social news site, the possibilities of sales promotion and conversion are endless.

The Big Players in Social News

If you've read this far and are ready to dive into the social news space, you will initially have to acquaint yourself with the multitude of social news sites that exist, as each one is different. You may want to visit some and skip others, and this is fine, since there are so many social news sites for just about any topic. The general social news sites are naturally likely to bring more traffic, so you should have a brief understanding of the more popular ones and be able to use them, especially if you are interested in dabbling in the social news space. Topical sites are good to explore as well, but you may not necessarily need to use all of them.

General Social News Sites

Social news sites are divided into two main categories: general social news sites that have a varied focus, and niche social news sites that are geared toward a specific interest. In this section, we review the more mainstream general social news sites.

Digg

Digg (*http://digg.com*) is by far the biggest social news site on the Internet, with over 34 million unique visitors per month.[3] Initially launched as a technology social news site, Digg (Figure 10-13) aimed to reach a more diverse population in early 2008, and is now a more general-focus social news site.[4] Popular topics include technology (due to its strong group of early adopters), offbeat news (especially as it relates to funny pictures and videos), and politics. Digg also has categories for world and business news, entertainment news, gaming news, lifestyle news, and sports news.

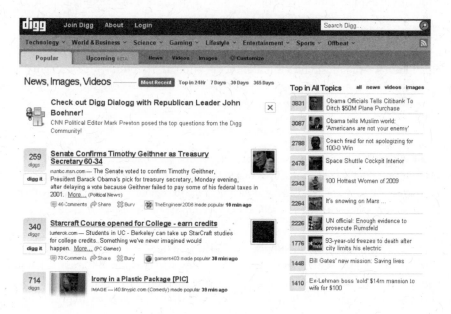

FIGURE 10-13. Digg's front page

Digg's dedicated pictures and video section is an incredibly useful asset to its members, though many would contend that the information contained therein is not "news," and that it is more appropriate to say that Digg is a social repository of interesting content.

On Digg, the popular jargon is:

- *Dugg*, for when a story is voted up.
- *Buried*, for when a story is voted down. Buries are not visible on Digg's page, but the algorithm accounts for all buries without making the information public.
- A *bury brigade* is a suspected group of individuals who bury any story that originates from a certain domain or that mentions a specific topic or individual.
- *Dupe* is a submission that may already be on Digg from another news source and was submitted by another member (competition is fierce on Digg, so don't dupe too many members if you can help it!).

To promote stories from users who are not your friends, Digg boasts a recommendation engine that suggests other users' submissions that are similar to the stories you may be Digging. Based on your Digging behavior, the service will propose similar stories submitted elsewhere on the service for you to explore and to vote on. Beyond the recommendation engine, each category has an "upcoming" page for you to view stories. Stories can also be sorted by most popular in the upcoming queue, and once a story is popular, it is removed from this queue. If you navigate, for example, to *http://www.digg.com/gaming/upcoming/most*, you will see the stories related to gaming that have the most votes but are not yet popular. Each individual category page has a dedicated upcoming queue, though you can view *all* stories on Digg's upcoming queue by navigating to *http://www.digg.com/all/upcoming/most*.

> **TIP**
>
> If one of your stories is sitting in a category's upcoming queue and there are several stories above it, give those stories a boost by voting them up! Once those stories hit front page fame, your story will have greater visibility at the top of the upcoming queue. By helping others in this regard, you're helping yourself. Remember, always be altruistic!

Digg has an ardent community of supporters and strong outside communities, though its management and support staff take no risks when they see questionable content submitted to sites. Users have been banned without recourse, and domains have been banned from being resubmitted as well. Digg's staff and users understand that marketers try to proliferate the site with self-promotional material without profit to Digg, so caution is strongly encouraged. Always tread carefully on the social space to avoid getting hurt.

reddit

The second most popular social news site is reddit (*http://www.reddit.com*). reddit has a strong group of supporters and a very active community. Unlike other social news sites, however, it (Figure 10-14) allows you only to submit a title to the site; there is no field for descriptions. Therefore, for success on reddit, it's imperative to have a descriptive title.

FIGURE 10-14. The front page of social news site reddit

reddit's terminology is simple: an upvote refers to a positive vote on a story, and a downvote refers to a negative vote on a story. The number of upvotes and downvotes is visible on the submission page, so you can find out exactly how well or how poorly your submission is performing.

TIP

When you use reddit and want people to vote on your stories, never link them directly to the submission page. The system has antispam measures in play, so if you link directly to the actual submission page and the story is voted on, the antispam measure may end up adding a downvote for that upvote. Therefore, your story may get no visibility (though it may appear to be performing well). Instead, link your users to a more public-facing page with numerous submissions and ask them to vote on several stories on that page so that your voting behavior looks as natural as possible.

Additionally, reddit operates on a karma system. If you get a lot of upvotes, you'll get more karma. This number fluctuates based on downvotes as well, so if you keep submitting high-quality content to the site, you should see a steady increase of karma.

Fortunately, if you submit a story on reddit and it does not perform well, you can delete it. You can then resubmit it at a better time or rewrite your title if necessary.

Another feature of reddit is the presence of subreddits, which are user-generated categories that let users submit their own stories. You can subscribe to any type of content on the site, from the generic subjects such as politics and technology to really granular topics such as home improvement, Firefox, bicycling, and even medical marijuana. These niche subreddits are not necessarily very popular or heavily trafficked, so visibility is not guaranteed.

NOTE

Because reddit is open source, you can map subreddits to your own domain and even style them to your liking. Niche site WeHeartGossip, explained later in this chapter, is a subreddit site that acts as a standalone celebrity gossip site.

Mixx

Mixx (*http://www.mixx.com*) is the third most popular social news site. On Mixx (Figure 10-15), you can submit a photo, video, or news article, and then you can assign it to up to eight categories. You can also tag submissions so that they show up on searches.

FIGURE 10-15. Mixx's popular articles across the entire site

The main feature that distinguishes Mixx from other social news sites is that it is a lot more personalized. Your home page is typically "Your Mixx," which is a collection of submitted stories in sections you have expressed interest in. Therefore, everyone's popular pages are customized based on their own interests. There are also hundreds of communities that have been created by users, similar to reddit's famous subreddits. You can't join every community you're interested in, unfortunately, as some are invite-only.

As an incentive for using the service, community members receive special awards based on their activities on the site. The biggest honor for Mixxers is the Supper Mixxer award, which is based on accumulation of karma points and solid commenting and submitting behavior. Comments are incentivized with "kudos" by members. Additionally, badges can appear on submitters' profiles to honor top submitters, conversation starters, thought leaders, most contentious submitters, hyperactive voters, newshound, and more. All of these awards are then stored in a user's Trophy Case.

One interesting thing about Mixx is that some users can flag a submission as "Breaking News," which makes it more visible to active members on the site. To ensure this tool is not gamed, only highly respected Super Mixxers can flag Breaking News.

The downside of Mixx is that while its community is extremely active and close-knit (in fact, there's an unofficial podcast by members highlighting great stories and users on the service), there is no real public-facing page of popular stories. Rather, individuals all see different popular pages based on their profiles and interests. Unlike reddit and Digg, which can claim thousands of unique visitors on promoted stories, Mixx popular stories receive only about 100 unique visitors. Therefore, it's a great site for community building, but it is not one that you'd want to focus on for social media marketing goals exclusively.

While Mixx has some shortcomings, its support team provides excellent customer service and always aims to satisfy its members. The team is extremely responsive and exemplifies the concept of "participation is marketing" (Chapter 4).

Shoutwire

Shoutwire (*http://shoutwire.com*) is one of the lesser-known but still popular social news sites. On Shoutwire (Figure 10-16), topics are general and front page fame is easy to achieve, though the number of unique page views is considerably low. A positive vote on Shoutwire is called a *shout* and a negative vote is called a *bash*.

FIGURE 10-16. Shoutwire's popular page

Similar to other social news sites, Shoutwire divides its submissions by article, picture, or video. In fact, as shown in Figure 10-16, the service even highlights very popular pictures and gives them great visibility. Therefore, a clever photo or memorable image can do well on Shoutwire.

Success on Shoutwire varies, but there are some ardent supporters who find that it can bring a few thousand unique page views, which is surprising given the site's reach of less than 20,000 monthly U.S. visitors.[5] The biggest challenge, then, is to find individuals who actually maintain accounts on this site.

NOTE

If you are excited about the opportunity to use the lesser-known social news sites but can't find individuals who have accounts on them, network onsite to find the community members and aim to get a diverse group of individuals to vote on your story. Never use the same IP address when voting on a social news site (or social bookmarking site StumbleUpon, for that matter). You may be banned for gaming the system. If you're at an office where your IP address is shared, vote from home. Don't risk your account credibility by voting on a story from the same IP address! This recommendation holds true for popular social news sites as well; Digg will also ban IP addresses used for gaming. Additionally, some sites have algorithms that will diminish the impact of a vote made by another user from the same IP address.

Propeller

Formerly known as Netscape (though perhaps not the one you're familiar with ever since it became a social news site), Propeller (*http://www.propeller.com*) was once a lot more popular than it is today, but has since been replaced by more mainstream social news sites. Once an active site with millions of unique monthly page views, Propeller (Figure 10-17) now receives less than 1 million page views a month.

Instead of having a voting system where every single vote counts as a point, Propeller works by ranking entries on a scale from 1 to 10, which accounts for various levels of user participation (comments, voting, number of times the story was physically accessed, and likely, participation on other parts of the site). A positive vote on the site is a *prop*.

In the past, Propeller had more categories, but the main categories on the service at this time are News, Arts & Entertainment, Business & Finance, Family, Humor, Sports, Style, Health & Fitness, and Politics.

Fark

Fark (*http://www.fark.com*) is a social news site that allows you to submit "news that is really not news,"[6] and anything goes on the site. The most popular stories typically are of the bizarre, odd, offbeat, and zany variety. In fact, front page stories can see tens to hundreds of thousands of views.

Fark (Figure 10-18) doesn't work on the same type of voting system as the other social news sites. In fact, only paying Fark members (a membership costs $5 a month) are allowed to vote on submissions. Additionally, safeguards are in play so that spam is not added on the site. All submissions need to be approved by a Fark team member before being published.

Yahoo! Buzz

There's a high probability that you already have an account on Yahoo!'s biggest social news site, Yahoo! Buzz (*http://buzz.yahoo.com*). That's because you can log in from any Yahoo!

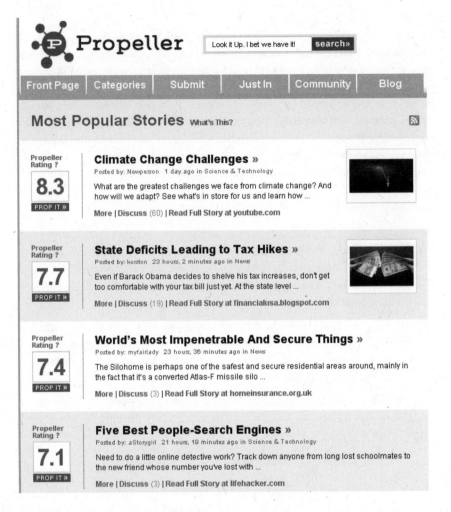

FIGURE 10-17. Propeller ranks entries on a scale of 1 to 10

property, making the site highly visible and accessible to Yahoo! users. This means hundreds of thousands of unique visits to stories that have hit the front page of the site.

Due to the mainstream focus of Yahoo! Buzz (Figure 10-19), the popular stories typically involve breaking news (political and national news) and entertainment. Your celebrity gossip story may have a lot of potential, so get as many people to *Buzz Up* the story as possible.

Yahoo! Buzz has the potential to bring millions of visitors to your site, so it is not a web property to be overlooked. The same rules of content creation and community engagement found on other news sites also apply on this site.

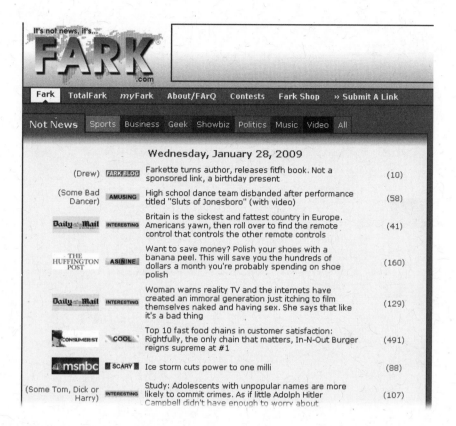

FIGURE 10-18. Fark's front page shows no votes at all

socialmedian

socialmedian (*http://www.socialmedian.com*) provides news that is relevant to your interests. The idea behind the site (Figure 10-20) is that your contacts, known as *newsmakers*, guide you through the most interesting content on the Web.

socialmedian lets you associate yourself with topics you're interested in, such as technology or politics. You can then start clipping stories that fit within your interests to share with the rest of the world and engage in an intellectual discussion with other individuals who may have clipped the story. socialmedian's success lies in its categorization, as its capability to place submissions into targeted categories is one of the most intelligent features among all social networks. The users of the site are also much more mature and serious-minded, and provide some great insights and feedback in comparison to other social news sites.

socialmedian has its own bookmarklet, which is handy if you plan to use the site regularly, and it also notifies you via email of popular stories in the topics you choose so that you can keep abreast of the latest happenings in subjects of interest to you.

FIGURE 10-19. The front page of Yahoo! Buzz

Topical Social News Sites

Surely, you're scratching your head at this point wondering how, with all the general social news sites, there could possibly also be topic-specific social news sites. There are! For those interested in a specific topic, these lesser-trafficked sites function in a manner similar to how tags operate in social bookmarking sites (see Chapter 9). These niche social news sites also boast active communities and great content. Note that only a few of the more popular niche social news sites are covered in this section.

kirtsy

kirtsy (*http://www.kirtsy.com*) is a social news site that covers women's issues in Arts & Entertainment, Design & Crafts, Family & Parenting, Fashion & Style, Food & Home, Internet & Technology, Mind, Body & Spirit, Travel & Leisure, Politics, and World & Business. Since the site is so focused on women, however, the popular sections of other social media sites (Internet and Politics) often do not get that much traffic on kirtsy (Figure 10-21). However, women's issues are extremely prevalent on the site, as are topics related to women who blog (as many kirtsy users do maintain blogs and the like).

FIGURE 10-20. socialmedian delivers personalized news directly to your home page

A useful feature on kirtsy is a section called the Editor's Picks, which allows hand-selected editors (chosen by kirtsy staff) to choose stories to be promoted to the front page of the site.

kirtsydoes not have a typical voting button. A vote is counted when you physically visit a web page. By visiting a web page, you are voting for the story that linked to it; other social sites do not require this (though their algorithms may account for visits and "blind votes," that is, votes that were made without the voter actually visiting the site). This tactic encourages you to visit the originating site (and you should be doing that anyway!).

If you click on a link and decide not to endorse the content, you can click on the Un-vote flower next to any entry. You can also add your negative vote by clicking Lose It.

Tip'd

If finance tickles your fancy, Tip'd (*http://tipd.com*) may be a great topical financial news site for you to explore. Topics range from personal finance to the economy to investing (Figure 10-22). When you vote on a story, you're *tipping* it, and you can also *topple* it to mark it as spam.

The site is relatively new, having launched in Q4 of 2008, but already brings more than 100 unique page views per front page submission and has some strong, intelligent, and engaging discussion. There are active moderators on duty who keep the site clean of spam submissions.

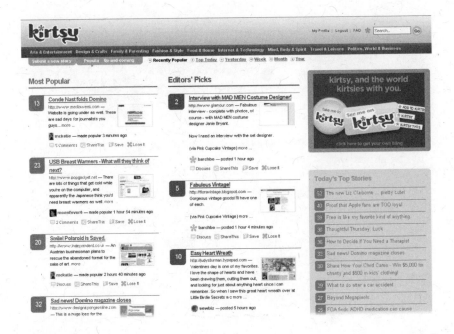

FIGURE 10-21. kirtsy's front page is divided into user-chosen stories and stories that were hand-picked by editors

Tip'd also integrates an interesting (though unrelated to social news) feature called SocialTickers, a real-time social financial news aggregator that captures public sentiment about a variety of popular stock ticker symbols using data from sources such as Technorati and Twitter. With this April 2009 development and the success of the site, Tip'd is poised to become increasingly successful in the upcoming months.

Sphinn

Social media marketing has a social news site of its own: Sphinn (*http://sphinn.com*). Sphinn (pronounced "spin"), shown in Figure 10-23, covers all aspects of Internet marketing and search, so stories about Google and Yahoo!'s successes or missteps are fair game, as are tips on how to maximize your Microsoft adCenter campaigns and even how to be a valued Sphinn contributor.

Like other social news sites, Sphinn prefers social news and informational articles. If you submit good content and are an active community member who votes on stories and adds valuable comments, you can do very well on the site.

Sphinn has a group of around-the-clock moderators who monitor the site for spam submissions and also keep the site clean. Sphinn also has a very active group of contributors and commenters; the comments are insightful and of high quality.

FIGURE 10-22. The front page of financial site Tip'd boasts stories, tags, and comments

When you *sphinn* a post, you're giving it a positive vote. Your negative vote on Sphinn is called a *desphinn,* and not only are negative votes public, but the individual desphinning the submission has to provide a reason why he/she does not endorse the content. All the same, the desphinn commentary is also rather insightful and informative and can often trickle into the regular comments section of the site.

TIP

Don't make your desphinn reasons personal. Sphinn is a professional Internet marketing social news community. At the same time, don't assume that a desphinn is a personal attack on the submitter or writer of the article. If your content is "desphunn," take it as feedback and let it be. If you're afraid of the criticism, it may be best not to submit to Sphinn at all, as your content can be easily scrutinized by members on the site.

FIGURE 10-23. The front page of Sphinn lets you view popular stories, find upcoming stories, and read comments

After reading this book, if you want to learn more and keep abreast of social media marketing developments, you should definitely sign up for an account on Sphinn so that you can get the most out of your newfound knowledge.

Slashdot

The Slashdot (*http://slashdot.org*) community predates many social news communities of today, but is still an incredibly popular social news site focused on technology (Figure 10-24). Slashdot is a social community for self-proclaimed "nerds."

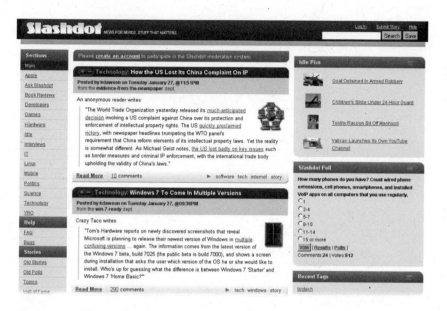

FIGURE 10-24. Slashdot always has ongoing discussions

The biggest part of Slashdot is the comments; they are scored on a scale of 1–5 (and can be labeled as Funny, Insightful, Interesting, and Informative).

Slashdot stories don't appear on the site unless approved by a site moderator. If you do have a good story related to technology, submit it to Slashdot; if it's really good, you may see a lot of unique visitors.

Hacker News

Hacker News (*http://news.ycombinator.com*), shown in Figure 10-25, is similar to Slashdot in focus with an emphasis on technology and science, but stories do not need moderator approval to appear on the site. Your submission will be accepted by default, though its visibility is determined by the number of upvotes it receives. The interface is most similar to reddit, though content cannot be downvoted, and as such, you will only see how many points (votes) the story received.

The Hacker News community is not too fond of Digg users; as such, the communities differ. Many stories do not get as much visibility on Hacker News as they would on a more mainstream website like Digg, despite its focus on technology.

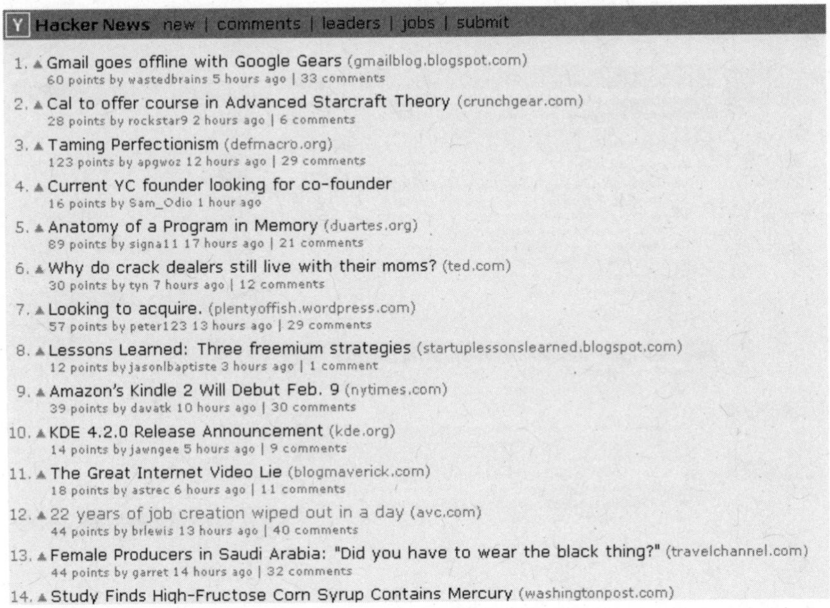

Y Hacker News new | comments | leaders | jobs | submit

1. ▲ Gmail goes offline with Google Gears (gmailblog.blogspot.com)
   60 points by wastedbrains 5 hours ago | 33 comments

2. ▲ Cal to offer course in Advanced Starcraft Theory (crunchgear.com)
   28 points by rockstar9 2 hours ago | 6 comments

3. ▲ Taming Perfectionism (defmacro.org)
   123 points by apgwoz 12 hours ago | 29 comments

4. ▲ Current YC founder looking for co-founder
   16 points by Sam_Odio 1 hour ago

5. ▲ Anatomy of a Program in Memory (duartes.org)
   89 points by signa11 17 hours ago | 21 comments

6. ▲ Why do crack dealers still live with their moms? (ted.com)
   30 points by tyn 7 hours ago | 12 comments

7. ▲ Looking to acquire. (plentyoffish.wordpress.com)
   57 points by peter123 13 hours ago | 29 comments

8. ▲ Lessons Learned: Three freemium strategies (startuplessonslearned.blogspot.com)
   12 points by jasonlbaptiste 3 hours ago | 1 comment

9. ▲ Amazon's Kindle 2 Will Debut Feb. 9 (nytimes.com)
   39 points by davatk 10 hours ago | 30 comments

10. ▲ KDE 4.2.0 Release Announcement (kde.org)
    14 points by jawngee 5 hours ago | 9 comments

11. ▲ The Great Internet Video Lie (blogmaverick.com)
    18 points by astrec 6 hours ago | 11 comments

12. ▲ 22 years of job creation wiped out in a day (avc.com)
    44 points by brlewis 13 hours ago | 40 comments

13. ▲ Female Producers in Saudi Arabia: "Did you have to wear the black thing?" (travelchannel.com)
    44 points by garret 14 hours ago | 32 comments

14. ▲ Study Finds High-Fructose Corn Syrup Contains Mercury (washingtonpost.com)

FIGURE 10-25. The Hacker News front page is extremely clean and simple to use

Small Business Brief

If you're running a small business, check out Small Business Brief (*http://www*
.smallbusinessbrief.com), a small niche community for news related to small-business
operations (Figure 10-26). On the site, positive votes are *fetched* and negative votes are *buried*.

Small Business Brief highlights its top users, which incentivizes its users to become active and
give to the community. Additionally, top sources are displayed on the site's righthand sidebar,
so high-quality submissions from domains can be acknowledged even further.

WeHeartGossip

WeHeartGossip (*http://www.weheartgossip.com*) was formerly known as Lipstick. Today, it
looks just like a standalone site, though it is actually a subreddit! Therefore, your reddit karma
translates over to WeHeartGossip (Figure 10-27). The focus of the site is celebrity news and
gossip as well as controversial stories.

FIGURE 10-26. The front page of Small Business Brief

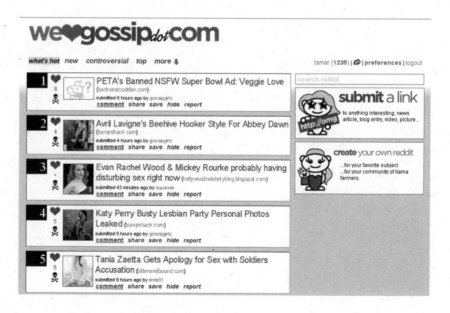

FIGURE 10-27. WeHeartGossip's front page: it doesn't really look like reddit

Since WeHeartGossip is a subreddit, the same rules apply for it as for reddit.

DesignFloat

The most popular social news website catering to web developers and graphic designers is DesignFloat (*http://www.designfloat.com*), shown in Figure 10-28.

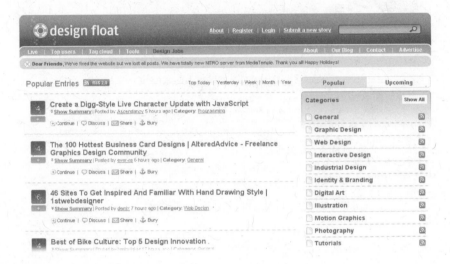

FIGURE 10-28. DesignFloat's front page is easily navigable

Care2

The biggest social news site for environmental and green news is Care2 (*http://www.care2 .com/news*), an active community for those who are environmentally and socially conscious (Figure 10-29). Topics focus on Environment, U.S. Politics and Government, World, Business, Green Lifestyle, Society and Culture, Animals, Health and Wellness, Science and Tech, and Offbeat news. The most popular topics on the site, however, are focused on the betterment of the environment through green initiatives.

The Care2 social network is part of the larger Care2 website, which is a general social network that empowers individuals to make a difference.

FIGURE 10-29. From the front page of the Care2 site, it is evident that emphasis is placed on environment and green news

The Pros and Cons of Social Bookmarking and Social News

In Chapter 9, you learned about social bookmarking. In this chapter, you learned about social news. Which method is better?

The best answer is both. When it comes to social news promotion, spread out your submissions as much as possible on a variety of social sites. While social bookmarking sites are more lenient about the kinds of content they will accept, by now you should know which content will perform best across a variety of social networks. If in doubt, study the communities you intend to target for maximum success.

Social bookmarking is generally used to index useful material that you can reference again and again. If you get enough bookmarks and your link is promoted to a popular section of the site, you may get those page views, but the tried-and-true motivations behind traditional bookmarking are echoed on social bookmark sites: people are simply storing the pages into their accounts. Therefore, you'll see visits from social bookmark sites, but you'll see far more visits on social news sites. On social bookmarking sites, you may see hundreds of users bookmarking your page, but only a fraction of them will actually visit it.

On the other hand, social news users know that they are getting timely content that could be outdated in a day or so. Therefore, you're likely to see a lot of page views in rapid succession on a social news site.

While some social news sites accept tags, they are not heavily referenced. In fact, stories on social news sites usually fall into oblivion after the front page visibility expires. In an experiment conducted by blogger Darren Rowse, it was found that social bookmarking services bring more long-tail traffic (that is, traffic over a longer period of time), but social news visibility was significantly reduced after its prime.[7]

Summary

Social news sites promote content that is endorsed by popular vote. These sites can bring hundreds or thousands of visitors to your site in a short period of time.

Originally, social news sites operated on the premise of the "wisdom of crowds," in which purely democratic voting would determine success. However, this concept has been disputed due to "herd mentality," in which certain groups of influencers can sway the entire public vote. It's become obvious over time that since these social news sites are really social in nature, content on social news sites can be manipulated. Many social news sites have safeguards to prevent against this type of gaming.

Getting to the front page of a social news site can be easy or difficult depending on the site. Factors affecting promotion include the user and his credibility on the service, the diversity of voters, the number of votes, the speed of votes, the category the story is submitted to, the number of negative votes, and the number of comments received.

You might want to be one of those users with high credibility and respect. If so, follow the 10 commandments to become a power user, but keep in mind that this will take time. An avatar is important, but your contributions and regular activity are also substantial factors. Networking is vital, and there are numerous ways to network both onsite and offsite, from voting on stories before they actually get promoted to the front page to adding witty and eccentric comments that make people smile.

The right kind of content is also important for social news mastery. People love lists, games, breaking news, controversy, videos, and pictures. All submissions should be accompanied with a good title (and description, if applicable). If you have a great piece of content but a bad title, people won't read your submission and you may not have another chance to succeed on a social news site.

There are a variety of general-interest social news sites. Digg is by far the largest, and reddit is close behind. Mixx, Shoutwire, Propeller, Fark, Yahoo! Buzz, and socialmedian are also important to know about, since each has its pros and cons.

There are an even greater number of topical social news sites. kirtsyis tailored to women, Tip'd is for the finance aficionado, Sphinn is a community of Internet marketers, and Slashdot and Hacker News are for fans of technology. WeHeartGossip is a gossip-centric subreddit that looks like a standalone site. Design Float is a great community for web designers and graphics professionals, and Care2 is for those who want to be environmentally aware.

Should you ignore social bookmarking and focus on social news? No. You should try both. Social news will bring more page views in a shorter amount of time, but social bookmarking is the gift that keeps on giving, as traffic doesn't die down as quickly.

Endnotes

1. *http://blog.digg.com/?p=73*

2. *http://blog.digg.com/?p=74*

3. *http://latimesblogs.latimes.com/technology/2009/01/kevin-rose-and.html*

4. *http://www.readwriteweb.com/archives/digg_the_decline_and_fall_of_tech.php*

5. *http://www.quantcast.com/shoutwire.com*

6. *http://www.fark.com/farq/about.shtml#What_is_Fark.3F*

7. *http://www.problogger.net/archives/2007/07/06/why-stumbleupon-sends-more-traffic-than-digg*

CHAPTER ELEVEN

New Media Tactics: Photography, Video, and Podcasting

By now, you've learned that social media is about the written word—for the most part. While a large percentage of the online community would rather read from a web page or whip out a book, there is an abundance of auditory and visual learners. Increasingly, video is starting to take a stronghold in our online society, and photographs always tell a great story. These alternative media tactics are rather successful, though not many individuals seem to dabble in the arts of video, photography, and podcasting.

Social media empowers amateur photographers and turns them into content creators. With tagging capabilities, commenting, embedding, and more, image-sharing sites are blossoming as social communities. Photography sites are also great implements for social media marketing, as they can help build awareness of your products. Evangelizing on these community sites can really add value to your business and marketing goals. Put simply, there are myriad opportunities to be gained by understanding these social media sites, the communities, and the features therein.

On the video and podcasting front, new media has enabled everyone to become a producer or director and even to have their 15 minutes of fame (and then some). With affordable technology for the home consumer, it has become easy for individuals to create their own podcasts using nothing more than a microphone. That's not the end of it, either: individuals can also grow their subscriber numbers to the thousands using multiple online tools that are free (or cheap enough not to break your budget). In the video sphere, after Google acquired YouTube in late 2006, it became evident that video was on an incline, and now that more and more people are engaging in video and watching the content, it is clear that Google's purchase decision was the right one.

In this chapter, we'll review these new media tactics for marketing gain. While perhaps these are not purely social media marketing methods, they are strategies utilizing newer technologies to engage the social community, and social sites that accept photographs and videos have empowered content creators to express themselves and to promote their own personal projects and products.

Using Your Pictures to Market Yourself

If you don't have a digital camera by now, why haven't you purchased one yet? Decent cameras can be had for as little as $30, though high-end consumer/professional ("prosumer") models are also available for as much as $4,500 or more. No matter how much or how little you spend on your camera, once your photographs are uploaded to a social network for photography, it doesn't matter how big your pockets are: you can still gain from social media marketing, as long as your photographs are of acceptable quality. Since most users of social media networks appreciate high-quality images, photos are ideal for marketing on photo-sharing social sites.

While there are many image-sharing websites, one clearly stands out above the rest for self-promotion. Flickr, once a simple photo-sharing site, is the largest image-sharing site and now accepts user-submitted video (up to 90 seconds long) as well. With billions of photographs stored and hundreds of thousands of paid subscribers (in addition to free users), Flickr is a powerful network packed with features. In terms of marketing, there are a number of ways to leverage the community features for personal gain. In this chapter, we'll focus primarily on Flickr for image sharing, as its functionality and community features are unrivaled by other social image-sharing sites.

Flickr: Your One-Stop Shop for Image Sharing

If you're looking to raise awareness about yourself or your products and to establish thought leadership in the creative arts, look no further than Flickr (Figure 11-1). Today, the Yahoo!-owned property has an incredibly large community with many followers and addicts, and as such, is a wonderful way to learn more about social engagement and perhaps social media marketing as well.

FIGURE 11-1. Flickr home page

TIP

On Flickr, overt marketing is against the community guidelines (*http://www.flickr.com/guidelines.gne*), which state that Flickr is not intended for commercial pursuits: "Flickr is for personal use only. If we find you selling products, services, or yourself through your photostream, we will terminate your account." Tread carefully if this is your sole goal for using Flickr. If community engagement and benefiting from the site's social features tickles your fancy, keep reading this section.

Signing up for a Flickr account

If you have a Yahoo! account, you can start by going to the Flickr page. While Flickr is part of the Yahoo! family of products, you'll still need to create a Flickr name to be recognized on the service (Figure 11-2).

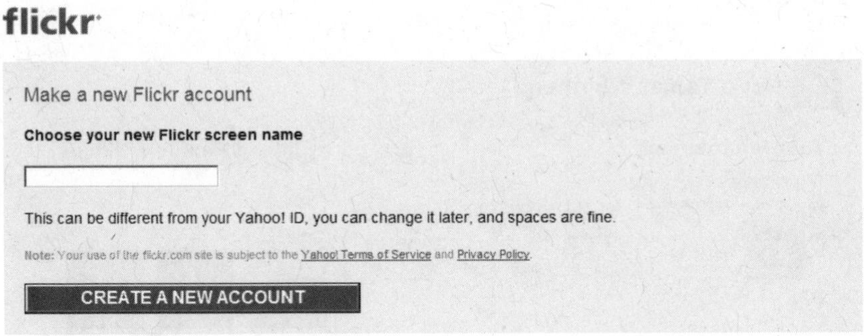

flickr·

Make a new Flickr account

Choose your new Flickr screen name

[   ]

This can be different from your Yahoo! ID, you can change it later, and spaces are fine.

Note: Your use of the flickr.com site is subject to the Yahoo! Terms of Service and Privacy Policy.

CREATE A NEW ACCOUNT

FIGURE 11-2. Creating your Flickr username

Once you create your name, you are asked to personalize your profile (Figure 11-3). First, you'll upload an avatar, otherwise known on Flickr as a *buddy icon*. After this, you'll choose a custom Flickr URL, which will become your personalized page for accessing your photos (*http://www.flickr.com/photos/yourname*). Bear in mind that you will not be allowed to change this URL once it's created, so choose your customized URL wisely. Then you'll be asked to provide identifying details: your first and last name, your time zone, your gender, and your relationship status. There is also a field where you can fill out information to describe yourself. This field accepts some HTML (such as links, bold/italic/underlines, tags, block quotes, and other basic HTML).

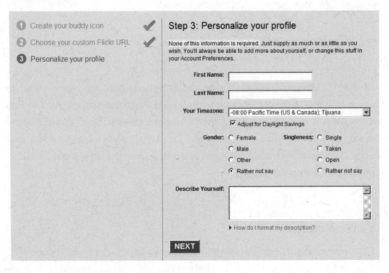

FIGURE 11-3. Personalizing your Flickr profile

After you've set up your profile, you can upload photos to the service (Figure 11-4). Since you're likely a free account holder at this point, you're limited to 100 MB of photos per month, so if you're using a high-end camera and have not resized or optimized the images using an external program, this 100 MB may run out quickly. Also, you can only store 200 photos on a free Flickr account at any given time, which may feel limiting, especially if you have a lot of photos to share.

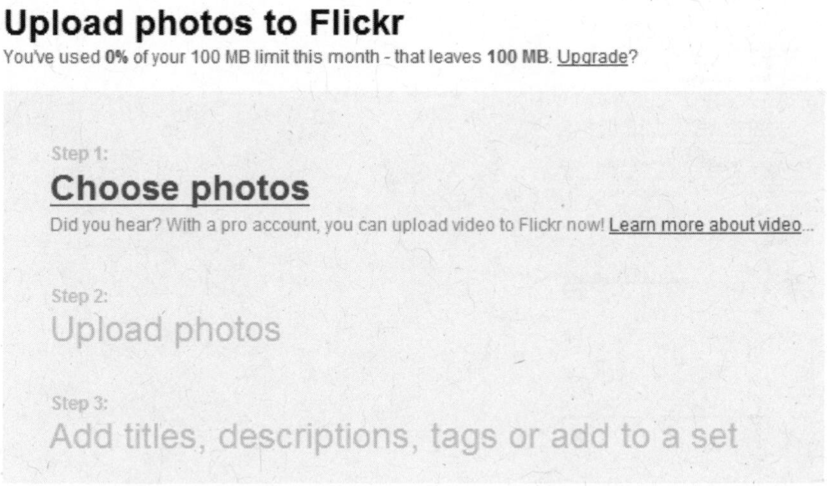

Upload photos to Flickr
You've used **0%** of your 100 MB limit this month - that leaves **100 MB**. Upgrade?

Step 1:
Choose photos
Did you hear? With a pro account, you can upload video to Flickr now! Learn more about video...

Step 2:
Upload photos

Step 3:
Add titles, descriptions, tags or add to a set

FIGURE 11-4. Choosing your first photos to upload to Flickr

NOTE
Flickr's paid account offerings, available at $24.95 for one year or $47.99 for two years, are highly desirable. You get unlimited storage for hundreds or thousands of uploads (if you'd ever need that). For more information about the perks of a paid Flickr account, see *http://www.flickr.com/help/limits/#28*. With a Flickr Pro account, you also get more authority; all paid members have a "Pro" badge next to their usernames, which gives them more credibility as active users of the site. Additionally, paid members can upload individual videos, but this feature is not available to free users.

With your first photographs uploaded, it's time to add a description that helps people understand exactly what you've just shared with the world (Figure 11-5). Limited HTML is allowed in description fields. You can also tag photos or even add a batch set of tags that will be applied to all photos that you recently uploaded. On Flickr, tags are *space-separated*, so your "new york city" tag may be better adequately named "nyc" or "newyorkcity" (dashes are not very popular on Flickr). Like delicious.com, commas that you enter will be added to the end of your tags, so avoid using them.

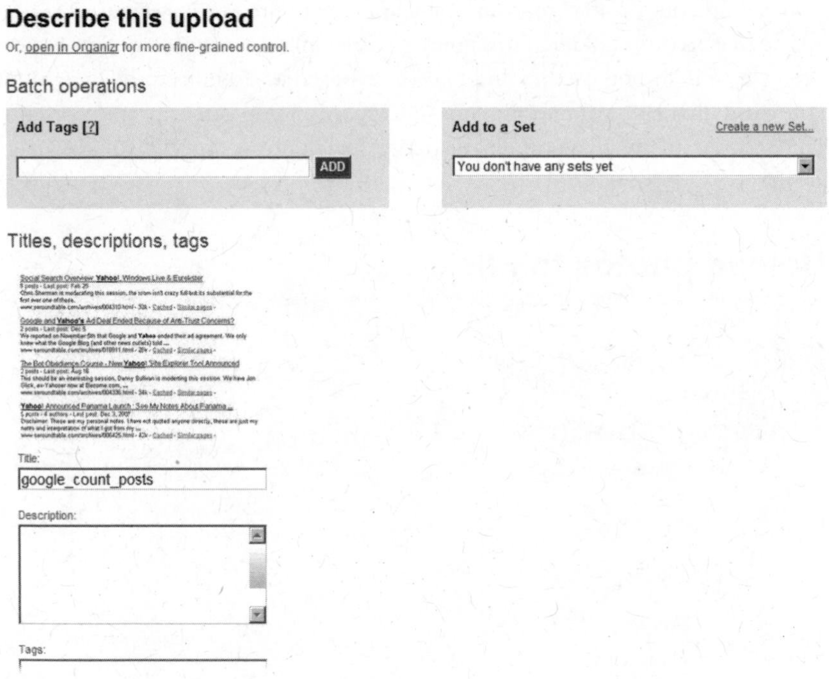

Describe this upload

Or, open in Organizr for more fine-grained control.

Batch operations

Add Tags [?]		Add to a Set	Create a new Set...
	ADD	You don't have any sets yet	▼

Titles, descriptions, tags

Title:

google_count_posts

Description:

Tags:

FIGURE 11-5. Adding a description to your Flickr upload

N O T E

In this chapter, we refer to Flickr features that are available to images. The same features are also available to video.

Tags are incredibly important on Flickr, so use them appropriately. If you want your images to be discovered, use the tags that the community would use. Thankfully, there is no limit to the number of tags you can add to a single image, though if you apply too many tags, you may be perceived as being "trigger happy." You can find popular tags on Flickr at *http://www.flickr .com/photos/tags*.

You can also add your photographs to a *set* (Figure 11-6). Sets are groups of photos. For personal purposes, you may want to upload your wedding photos into one set and the birth of your baby into another set. Similarly, you may want to dedicate a single set to your product uploads (just as long as you're not overtly selling them, in accordance with Flickr guidelines) and another set to a hosted community get-together. Organizing photos into sets lets viewers and visitors easily navigate and access your images.

The organization of sets within Flickr (Figure 11-7) may require a slight learning curve. The Batch Organize tool features a sophisticated Ajax interface that lets you drag and drop photos

FIGURE 11-6. A typical Flickr set

into a single container unit. You can also apply batch-editing functions to the photos or arrange photos in a set by specific criteria.

Flickr also has *collections* (Figure 11-8), available only to paid members, which are groupings of sets. Free memberships do not include access to collections and are limited to a total of only three sets. If you find yourself uploading many photos, you may want to have further control over your sets, so you'd choose to group them into collections. For example, your sets of get-togethers can be grouped into a collection of Flickr photos called "Our Events."

After you've uploaded your first photos on Flickr (and/or explored sets), it's time to take advantage of the community features. Flickr lets you add your friends via email (Yahoo! Mail, Gmail, and Windows Live Hotmail), but you can also search for users already on the service. If you're in the business world and not active in social media, it may be hard to find users you know on Flickr, so feel free to start with your social media contacts first. Of course, you can also start making friends by becoming active within the community, which is strongly encouraged.

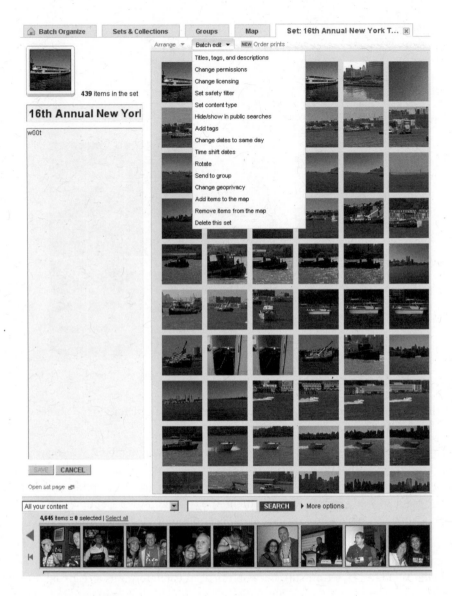

FIGURE 11-7. Organizing your Flickr set

Flickr's community features

You now have a basic understanding of how Flickr works, and it seems like any other photo-sharing site. However, there are a number of community features that you can make use of on Flickr. The most important feature is Flickr groups (*http://www.flickr.com/groups*). Groups are essentially sets that all members of the community can contribute to (whereas you are the

Beijing 2008

A collection of photos from Beijing, China during the Olympics in August, 2008.

Best of Beijing 2008
62 photos

Wangfujing Night Market
23 photos

Olympics - Table Tennis
60 photos

Olympic Green At Night
31 photos

Olympics - Swimming
41 photos

Olympics - International...
50 photos

Beijing Olympic Food Gallery
48 photos

Ogilvy Beijing Office
48 photos

Olympics - Beach Girls
36 photos

Olympics - Women's Beach...
83 photos

Kleenex "Let It Out"...
19 photos

Olympics - Men's Basketball
39 photos

Olympics - Men's Gymnastics...
80 photos

Bookworm Blogger Meetup -...
7 photos

Temple Of Heaven
56 photos

Panjiayuan

Beijing Subway

Lenovo at

Beijing

Beijing - Day 0

FIGURE 11-8. A typical Flickr collection

only one who can contribute to your own personal sets). Groups also feature discussions related to the topic matter.

The New York City group is shown in Figure 11-9. This group has nearly 14,000 members, and users have uploaded over 229,000 images related to the Big Apple. If you've visited the city, you can upload your photos to the group. If you own a restaurant or a small Mom & Pop shop in the city, you can share photographs of your business and happy customers therein. If you own a travel company specializing in New York City tours, you can upload photos of your customers' experiences or even your own as the official tour provider. Of course, to keep in accordance with Flickr's guidelines, do not promote your services heavily in the description of

FIGURE 11-9. The Flickr New York City group

the photograph. You should share related images that are still relevant to the community site at large; never be too self-promotional.

You can search for other existing groups related to your interests by navigating to *http://www.flickr.com/search/groups*.

When using groups, be advised that each individual group has distinct rules of engagement. You will want to keep your photo or video uploads relevant to the group to avoid having a moderator remove you from it.

Groups are important because the most active Flickr users are involved in them. As such, groups provide a great vehicle to build your network and to meet new people. If you join a group, be sure to actually contribute to it on a consistent or semiconsistent basis to become an established participant. Also, engage in conversation in the discussion forums where you can provide valuable responses and insights—or simply add comments on photos.

You can certainly choose not to use groups on Flickr, but this is the community feature most likely to give you exposure on the site. Of course, if you use a site that aggregates your social activity (such as FriendFeed, Figure 11-10), your activity on Flickr can be posted elsewhere, and other people will be able to access your photographs.

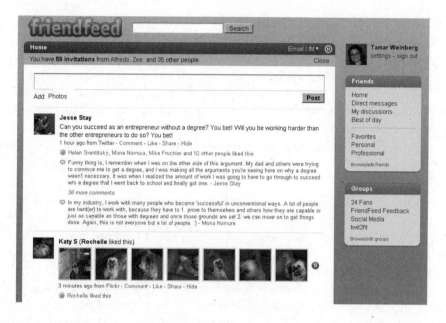

FIGURE 11-10. FriendFeed shows your social media activity, which includes Flickr

FRIENDFEED: AGGREGATING YOUR SOCIAL ACTIVITY

FriendFeed is a social site that aggregates social media activity around the Web. It may more appropriately be termed a "microblogging" application, but it is really so much more.

FriendFeed consolidates updates from numerous social media sites and services, including blogging platforms, social bookmarking sites, book-sharing sites, social news sites, status updating sites (Twitter, anyone?), social music sites, video-sharing sites, comment update sites, and other services, such as photo-sharing site Flickr and video-sharing site YouTube (to be discussed later in this chapter). When you create an account on FriendFeed, you indicate which social media sites you want FriendFeed to display. Your shared photos on Flickr, your Facebook status updates, and your

StumbleUpon reviews and submissions, in addition to any or all other services you add to your FriendFeed stream, will be accessible via a dedicated URL for your username, such as *http://www .friendfeed.com/tamar*. Users who subscribe to your FriendFeed feed will be able to see when you update new photos to Flickr, and when you do, your images can get even more exposure.

You may also choose to share relevant photos on other social sites, such as Twitter and delicious.com where applicable, though utilizing that option becomes more of a manual process than an automatic one.

Creative Commons: Licensing your images for sharing

Flickr allows you to specify your images as being under the Creative Commons (CC) license (*http://creativecommons.org*). A CC license turns your photos from "All Rights Reserved" into "Some Rights Reserved," which allows other individuals to share your photographs and images as long as you are credited as the original creator/photographer. In general, the minimum CC license is the Attribution license, which allows your images to be distributed, remixed, tweaked, and built upon under the condition that the individual reusing your image gives you credit. If you do not mind having your images reshared, and if you want to actively promote your creativity offsite, it's a good idea to upload your images to Flickr under the CC license for additional exposure.

If your photos are CC-licensed, you may still be contacted by bloggers and media outlets (newspapers, online magazines, and other publications) requesting permission to use your images offsite (in other cases, once you set your images to a CC license, users may reference those images in their blogs or news articles without contacting you first). One example of this type of media outreach is via NowPublic.com, a news-sharing community in which writers gather the majority of their news article images from Flickr. If your photos are a good fit for the network, you'll be direct-messaged by a member of the service asking you if you want to share your image. If you get credit for these images on alternative sites, you can use your Flickr leverage to gain credibility offsite as well.

The popularity of your pictures: Flickr's statistics

Are people enjoying your photos? Is it worth the effort or time investment? Find out by checking your Flickr user statistics (Figure 11-11) on a semiregular basis. Similar to an analytics package, these statistics give you insights into how many people are viewing your images at any given time in addition to gauging the most popular images in your photostream.

By noticing these trends and what is widely received by members of the site, you can determine which types of images or videos work best for the site's audience.

FIGURE 11-11. Flickr image statistics

"Marketing" on Flickr

So now you know a little about how to engage in the community and market your products on Flickr. There's more to be done, though.

Your profile page is a great place to start advertising yourself and your services. Use your profile page to talk about yourself and what you do. That way, if you become respected within your groups and within the community, people will get to know a little more about you. Again, don't overtly promote your business motives; Flickr is a social site and does not condone commercialism. However, if your biography says something along the lines of "My name is Tamar Weinberg. I love photography, New York City, and LOLcats, and I just wrote a book on social media marketing. Find out more about me on my blog at *http://www.techipedia.com*," you should be safe. After all, that's an overall picture of who you are and does not focus only on your professional and commercial endeavors.

You can also use your photostream to share high-quality images of the products and services you offer. Matt McGee, an avid Flickr user, suggests the following applications for Flickr:[1]

> A general contractor, for example, would upload photos of homes or commercial buildings you've built. A winery would upload photos of grape harvest, the winemaking process, and even your tasting room and close-ups of your wine bottles. A caterer would upload photos of events you've worked — your staff, the food, the overall presentation, etc. A woodworker would upload photos of items you've made, your workshop, etc.

McGee advises talking about your product in image descriptions, but cautions against the direct sales pitch. Of course, you should avoid promoting your product and service offerings in discussion groups as well. You can, however, link back to your original site, where you may be able to provide more information about that house you've constructed or the detailed winemaking process.

One of McGee's Flickr images is shown in Figure 11-12. A great page on Flickr has a detailed title (preferably with the date so that the image doesn't get filtered via a regular search, since duplicate or similar titles are often omitted or harder to find) and valuable description. McGee added the image to his personal set as well as to a number of user groups on the service. He also added tags to the image (as you can see, other users can add a pertinent tag to user images, too—just in case!).

Don't overlook the community features on Flickr, and do participate in the community. If you're sharing for the sole purpose of sharing, you may find that you can effectively market yourself without overtly doing so!

Other Photo-Sharing Contenders

Flickr may seem like the best choice given its community features and potential opportunities for marketing. However, there are other photo-sharing sites where can host your images.

SmugMug (http://smugmug.com)
> SmugMug is a popular choice for image sharing among professional photographers. A SmugMug account costs upward of $39.95/year.

PBase (http://pbase.com)
> The PBase photo community is great for professional photographers, but accounts are not free ($23/year for 500 MB of storage; $60/year for 1,500 MB of storage).

Photobucket (http://www.photobucket.com)
> Photobucket, an image-storing facility above all else, is owned by Fox Interactive Media. For the most part, it is used for personal photography. Photobucket owns TinyPic (*http://www.tinypic.com*), another photo-sharing site, which allows you to upload an image to a server and get an immediate URL to share with friends. Accounts on both Photobucket and TinyPic are free, and you do not need an account to utilize these URL-sharing features.

Picasa (http://www.picasa.com)

Picasa, a Google-owned file-sharing website, is free. It's not a popular site for social media promotion, but is favored by families who enjoy sharing photos of significant events and milestones with their loved ones.

Zooomr (http://www.zooomr.com)

Zooomr, the closest rival to Flickr, has a lot of features offered by Flickr, and boasts a growing and passionate, but much smaller, community. Like Flickr, Zooomr offers both free and pro accounts.

FIGURE 11-12. A Flickr Page that takes advantage of all the features

These sites all pack in image-sharing capabilities, but are not as exhaustive in terms of features and functionality as Flickr—at least for marketing and community. Zooomr comes closest in its offerings and functionality, while SmugMug and PBase are the preferred sites of professional photographers. If you want a no-risk approach to sharing your images, you may want to try one of the Zooomr or SmugMug paid accounts instead of Flickr.

Photoblogs and Galleries: Dedicated Portals for Images

Although not used heavily due to the existence of image-sharing sites, photoblogs (blogs that are intended for image sharing only) are a great place to share your images and to tell people about your professional or personal pursuits. Additionally, personally hosted image galleries are excellent ways to start promoting your photography or business-related images.

Like blogs, photoblogs and galleries require hosting for the files, a domain name registration, and possibly some technical knowledge. You can use standard blog applications (see Chapter 5) to manage your photographs, and they'll double as photoblogs, but keep in mind that if you host lots of images on your server, you may run out of space and incur penalty fees with your host. Be sure to keep your images small (that 4,500×6,000-resolution image can be reduced to half of that size or even less) and check your server storage space regularly, especially if image hosting is something you plan to maintain. This way, you won't run into problems when you run out of hosting space. Ideally, you should never delete images from your photoblog, because they can result in broken pages and links (404 errors). Instead, just keep the file sizes small.

HOTLINKING: THE COPYRIGHT INFRINGEMENT AND BANDWIDTH THEFT DILEMMA

Can your images be stolen? Unfortunately, yes. It's easy for people to right-click an image, get the URL, and link to it from their own sites. The problem with this tactic is that if too many people "hotlink" to your image, as this practice is called, you're providing your image to another site without explicit permission and can run out of bandwidth quickly on your host without even noticing. If other people start accessing your artwork from elsewhere on a regular basis, your bandwidth allocation can get sucked up quickly. Further, this method of redistributing artwork can be considered a copyright crime. Naturally, this is a risk you will have to take if you host images on a photoblog. With Flickr, there are safeguards to prevent users from stealing your images. When you're hosting your own images, it's a bit more difficult. Search the Internet for a script that will help you prevent others from hotlinking your images (that means that the images won't be linked from a site you didn't authorize). You can find a script that replaces a hotlinked image with a placeholder image of your choosing (such as "Original image taken from *mysite.com*"). Depending on your server configuration, the scripts may differ, so technical know-how is important. An alternative is to contact the person who used your image without permission and to inform her that she is stealing your bandwidth, in hopes that she will remove the image and cooperate with your request. Not all individuals who steal content are willing to comply, but it doesn't hurt to ask.

If you are limited in space, it's not a bad idea to use one of the other image-sharing sites and then link to the images on your blog or website. You may want to make sure that the images remain permanently stored on these sites if you end up using these photo-sharing web applications.

If you choose to use a photoblog application, in addition to the standard WordPress and MovableType hosted applications that are available (which can be reserved for your blog posts or dedicated to photography), you may want to try more dedicated photoblogging tools.

Pixelpost

One such hosted application is Pixelpost (*http://www.pixelpost.org*), which operates on the premise that you're going to be uploading a single image a day. You can change the daily photo rotation feature in the backend if you wish. Figure 11-13 shows an entry from a Pixelpost blog.

Gallery

One of the most frequently used tools for hosting images is Gallery (*http://gallery.menalto .com*). Gallery (Figure 11-14) touts itself as a photo album organizer with a surplus of features for security and photo management. It features support for comments with an optional CAPTCHA, email, and even offline uploading applications.

Coppermine

Coppermine (*http://coppermine-gallery.net*) is another frequently used photo gallery. If offers features similar to Gallery and includes support for slideshows, multimedia, and more (Figure 11-15).

Once you install a gallery application, you can tweak it to your liking; you may want to set up different albums for event photographs and for product photos. Gallery and Coppermine give you a lot of options to customize albums within the application and even to determine the gallery's layout, so do not feel restricted and have fun (however, for layouts, especially if you want them to match your own site, you may need to enlist the help of a savvy web developer).

Beyond Stationary: The Video Marketing Guide

Photographs are one thing, but video is becoming more popular by the minute. The online success of the streamed video of the Obama inauguration is a clear indication of the role video and streaming video are playing in this day and age. According to content delivery network Akamai,[2] the inauguration of the newest U.S. president set a record for the most simultaneous streaming video viewers in the Web's history. This rivals that of actual television.[3] In fact, video streaming of the inauguration was banned at my husband's workplace due to the possible bottleneck it would create for Internet access in his populated office. Instead, employees were confined to a conference room television. This all goes to show that online video is becoming

FIGURE 11-13. A Pixelpost-powered photoblog

more pervasive. I used to be reluctant to watch online video primarily due to its potential to disrupt my schedule, but I now find myself watching one or two videos a week (and sometimes even more), which is substantially more frequent than I'd have ever imagined. Not all content is accessible in the written word any longer. Video is here to stay.

You may recall reading about the success of Blendtec in Chapter 2. To refresh your memory, the company, which manufactures personal and commercial-grade blenders, started a channel on YouTube with a mere $50 marketing budget. By creating and sharing videos of people

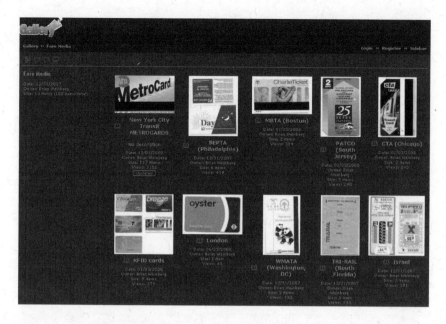

FIGURE 11-14. A Gallery-powered photo album

blending toys and electronics to show how robust its blenders were (and by awing people with these sometimes highly dangerous feats), Blendtec became one of the most pivotal blender manufacturers of all time, and probably the biggest presence among blender producers in the online arena.

Blendtec isn't alone. Product placement is working in these user-generated videos as well. One case in point is Fred (*http://www.youtube.com/user/Fred*), a YouTube sensation from the summer of 2008. Fourteen-year-old improv actor Lucas Cruikshank plays the part of Fred, a 6-year-old who has anger management issues, an alcoholic mother, and an imprisoned father. Using YouTube, he's turned himself into an instant celebrity and has amassed nearly 100 million video views across more than 30 videos (Figure 11-16). Fred's YouTube channel, in fact, is the #1 most subscribed YouTube channel of all time. Fred's popular videos include "Fred Goes Swimming," "Fred Loses His Meds," and "Fred's Mom is Missing," all of which have been viewed by millions of viewers and fans. After Cruikshank's initial success, he teamed up with a company called ZipIt Wireless[4] to target the younger demographic. In a number of videos, Fred can be seen using a ZipIt Wireless handheld device to communicate with his friends. Now, there are teens and tweens around the world who want to have the same device that Fred is using, albeit subtly, in these videos.

Another case that illustrates the benefits of product marketing is a video demonstrating the unboxing of the Samsung Omnia, a mobile phone.[5] In an unboxing video, you would expect to see a guy opening a box, finding a bunch of accessories, and then digging into the box to

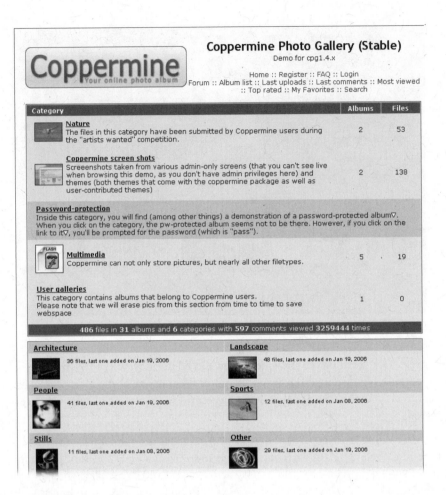

FIGURE 11-15. A Coppermine screenshot

take out his brand-new phone. That isn't the case for this video. When the man recording the video opens the box, he finds only a red button. The man presses the red button, and out of the box comes a marching band. Clearly, that isn't something you'd expect. The video has thousands of five-star ratings for its originality and resourcefulness.

With the right kind of message and a viral hook, videos can travel far. In fact, thousands of creative video artists have become superstars on video-sharing sites like YouTube. Alternative rock music group Weezer capitalized on this phenomenon with a unique twist in its "Pork and Beans" video.[6] In the video, celebrities who became popular from YouTube were brought together to film one of the most interesting and unique music videos of all time. If you have followed the memes in these individual YouTube videos, you can understand and appreciate the humor in the Weezer music video, and that is why it has been viewed more than 17 million

FIGURE 11-16. Fred's videos on YouTube

times since its release. The success of the music video shows that Weezer understood that there's an entirely separate Internet subculture and that it is a force to be reckoned with. Since there already exist hundreds of celebrities made popular simply by virtue of being featured in highly viral videos on YouTube, it's evident that the video-sharing site is one where people and products can be influencers.

Marketing on YouTube

The previous examples illustrate the fact that YouTube is a wonderful tool for product marketing, and that the right video can get hundreds of thousands or millions of views, fans, and of course, real conversions that matter. But how do you market your videos? Are there tips and tricks to get the most out of the video-sharing site? Absolutely!

But how, exactly, do you do it?

In the examples in this chapter and in others throughout the book, creativity wins out. If something's bizarre, zany, funny, informative, or just something completely unexpected, your marketing message can travel very far. You might even find that you're an unanticipated celebrity in the aftermath.

With more than 70 million U.S. views alone on social video-sharing sites like YouTube,[7] online video is not a marketing channel to be overlooked. A study performed by analysis firm eMarketer (Figure 11-17) found that 88% of all Internet users will be watching videos by 2012.[8]

US Online Video Viewers* and Online Video Advertising Viewers**, 2007-2012 (millions and % of Internet users)

	Online video viewers*	% Internet users	Online video advertising viewers**	% Internet users
2007	137.5	73.6%	111.4	59.6%
2008	154.2	80.0%	129.5	67.2%
2009	167.5	84.1%	144.1	72.3%
2010	176.0	85.7%	154.9	75.5%
2011	183.0	86.8%	164.7	78.1%
2012	190.0	88.0%	174.8	80.9%

Note: at least once per month; *downloads or streams video (content or advertising); **views any form of video advertising (in-stream, in-banner, in-text)
Source: eMarketer, November 2008

099521   www.eMarketer.com

FIGURE 11-17. The number of U.S. online video viewers and online advertising viewers from 2007 to 2012

In fact, comScore reports that 14.5 billion videos were watched in March 2009 alone[9] across 10 different video and video-sharing sites (Figure 11-18). Not surprisingly, nearly 6 billion of these video views originated from YouTube, which is one of the top three most popular sites on the Internet, according to Alexa.[10]

Convinced? Let's see how you can market on YouTube.

Creating your video

First, let's take a look at considerations for your video before you actually upload it to a video-sharing site.

- Your equipment does not need to be costly. Prolific video blogger Loren Feldman uses a $129 Casio point-and-shoot camera for the majority of his videos (and in the remaining cases, he uses the camera that came with his MacBook).

- Your video shouldn't be too lengthy. Your viewers have short attention spans, and unlike with print media, they can't multitask with video. Two to three minutes is the sweet spot. If you can capture the essence of your video in a shorter timeframe, you are encouraged to do so.

Top U.S. Online Video Properties* by Videos Viewed November 2008 Total U.S. – Home/Work/University Locations Source: comScore Video Metrix		
Property	Videos (000)	Share (%) of Videos
Total Internet	12,677,063	100.0
Google Sites	5,107,302	40.3
Fox Interactive Media	439,091	3.5
Viacom Digital	324,903	2.6
Yahoo! Sites	304,331	2.4
Microsoft Sites	296,285	2.3
Hulu	226,540	1.8
Turner Network	214,709	1.7
Disney Online	137,165	1.1
AOL LLC	115,306	0.9
ESPN	95,622	0.8

*Rankings based on video content sites; excludes video server networks. Online video includes both streaming and progressive download video.

FIGURE 11-18. Over 14 billion videos were watched in March 2009

- Steer away from the traditional advertising mindset. Today, people fast-forward through commercials because generally commercials overtly market some sort of product. Don't use user-generated content mediums to promote yet another advertisement. You'd expect the Samsung Omnia unboxing presentation to show a smartphone. Did watchers expect a magnificent marching band performance? Certainly not. However, the video evoked an emotion of wonder and amazement, and the ad itself cleverly made individuals aware of that particular mobile handset (and possibly motivated them to buy one, too, in hopes of finding a marching band inside the box).

- Where possible and if necessary, embed your business URL into the video. Blendtec's branding is apparent on each of its individual "Will it Blend?" episodes, both before and after the videos.

Tips to promote your video on YouTube

One of the most coveted spaces on YouTube is the service's Popular page (*http://www.youtube .com/browse*). Having this front page visibility can guarantee thousands of simultaneous views, and unlike on other social sites, you won't have to be responsible for the traffic surge's impact on your website hosting contract! But how can you get your videos to the forefront? How can you have your videos promoted to the top? It will take a lot of dedication and effort. Most importantly, you're not going to be able to come out of nowhere and hit the front page unless you meet the needs of the community and produce a video that raises eyebrows.

On average, frequent activity on YouTube will help contribute to your overall YouTube promotional efforts. Do you have friends on the network? Does your channel have subscribers? Do your videos actually get views? If you have credibility on the service, the likelihood that your brand-new uploaded viral video will become successful is higher than it is for the same submission from someone with no prior activity or strong account standing on the service. Therefore, if you're considering a viral marketing campaign in the video arena, participate first. Share some small-scale videos initially. Be a community participant. Don't come out of nowhere and expect instant fame; this will just raise suspicion.

If you want an already uploaded video to become successful on the video sharing site, think again. Like most social sites, there's a period on YouTube during which newly uploaded videos have more influence. If videos don't gain initial traction within the first 48 hours of being uploaded, they may never become popular. However, if you have a great video that you feel deserves the community's attention, you may want to upload it again to the service and retry.

Be sure that when you do upload your video, you add it to the appropriate category. A good description and appropriate tags are also important (Figure 11-19).

> **TIP**
>
> If you plan on uploading more than one video (as in Fred's case or in the Blendtec series of videos), you may want to add unique tags specific to your uploads so that YouTube's "Related Videos" section will show your other videos. If your tags are too generic and common among other YouTube videos, your visitors may not find your content thereafter—they'll likely find related videos from other publishers instead, unless they navigate directly to your YouTube channel (profile) page.

Once the video is uploaded and processed, it's time to start gathering those unique views, comments, and ratings. If you can share the video with your network and leverage your social media community for votes, you can kick-start your video's success. Moreover, on almost all video-sharing sites, you don't have to limit your viewers to the actual video page; as the content producer, you can also enable embedding of videos so that other bloggers and video producers can share your videos on their own blogs or websites. This is another way to encourage and facilitate sharing that isn't limited to the actual video-sharing site.

Getting those video views, ratings, and comments is especially important within the first few hours. Share the videos as much as possible with your friends and colleagues. If you already have an established network on YouTube, you can start by sharing with the people who already follow you. Of course, you can leverage social bookmarking and social news sites as well; you may recall that numerous social news sites have dedicated video sections that can provide increased chances at visibility.

If visibility is important, obtaining critical mass is vital. Reach out to bloggers who write on topics relevant to your video content. Do the same on forums and discussion boards and other social sites, including Facebook and MySpace.

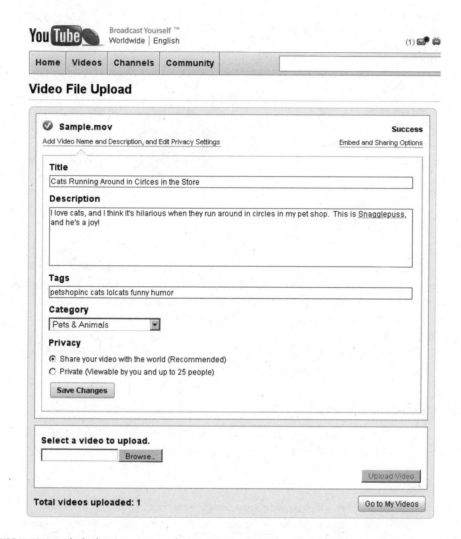

FIGURE 11-19. A good title, description, tags, and appropriate categorization are essential for YouTube videos

Find out how your video is performing

Who is actually watching your video? Where are viewers coming from? How popular is the video? Once your video is uploaded to YouTube, you can view detailed statistics via the YouTube Insight tool (Figure 11-20), which breaks down video performance by views, discovery (referrals), demographics, and hot spots (popular points of viewership within the video itself).

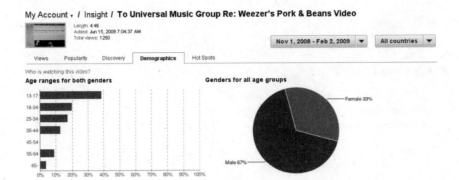

FIGURE 11-20. YouTube Insight provides demographic information

Using YouTube Insight, you can discover exactly who is watching your videos, in addition to the waxing and waning periods of your videos' popularity. You may find out that your videos have an unexpected following in a foreign country, or that your product video is appealing to the parents of the kids you're actually trying to target. You may be surprised by the data and make unexpected discoveries, but if you regularly contribute video content on YouTube, this information can arm you with the figures you need to create more compelling videos in the future and to schedule or target them more effectively.

Other Video-Sharing Sites

YouTube is really where most of the action is. Since it has the most followers (almost 6 billion videos viewed monthly? Yes!), it's a property not to ignore.

> **N O T E**
>
> The most popular online video properties are Google sites, which include YouTube and Google Video. The latter is not as heavily trafficked. Additionally, several of the sites listed in the chart in Figure 11-18 are not video-sharing sites. For example, Hulu streams television from popular cable networks such as FOX and NBC, but does not accept user-generated content.

There are several other sites where you should consider sharing your videos. These include:

- General-purpose video-sharing sites, such as Blip.tv (*http://blip.tv*), DailyMotion (*http://www.dailymotion.com*), Google Video (*http://video.google.com*), Yahoo! Video (*http://video.yahoo.com*), MySpace Video (*http://vids.myspace.com*), MetaCafe (*http://www.metacafe.com*), Revver (*http://www.revver.com*), Spike (*http://www.spike.com*), Veoh (*http://www.veoh.com*), Viddler (*http://www.viddler.com*), and Vimeo (*http://vimeo.com*).

- How-to and instructional video sites, such as 5min (*http://www.5min.com*), Howcast (*http://www.howcast.com*), and Sclipo (*http://sclipo.com*).
- Comedy video-sharing sites, such as Break (*http://www.break.com*).

If you're interested in detailed metrics for videos, check out TubeMogul (*http://www .tubemogul.com*) (Figure 11-21), which aggregates video-viewing data from a multitude of video-sharing sites (more than 20, including many of the sites listed previously) so that publishers and content producers can understand how viewers are engaging with video. The basic version of TubeMogul is free, though for heavy business use, paid options are available with additional features (*http://www.tubemogul.com/about/features.php*).

FIGURE 11-21. TubeMogul shows detailed video analytics

Yes, YouTube is the most viable choice for video sharing, but don't discount the smaller online properties. If you don't upload that video to another site, someone else may use a program to download the video off YouTube and redistribute it on those other video-sharing properties for you (which you can usually get removed if you own the copyright to the original video). If you want complete control over statistics and want to delete comments that may be questionable, be the first person to upload the video on those individual websites.

The Art of Videoblogging

The written word is not as effective as face-to-face communication. This concept has given rise to videoblogging, which has become a successful marketing tactic for the affable and sociable individual. Gary Vaynerchuk has personified this idea with Wine Library TV, a several-times-a-week video blog in which Vaynerchuk, a wine connoisseur, makes wine more accessible to the general public. Born into his family wine store business, Vaynerchuk spent his college years with an entrepreneurial mindset and established himself as a wine expert by trying out every single wine sold in his family store. Today, his Wine Library TV blog (*http://tv.winelibrary .com*) is the most successful channel specializing in wine on the Internet. His family business, once netting $4 million a year, now rakes in more than $50 million of yearly profit. His charisma and ebullient personality have brought his channel nearly 100,000 regular viewers.

Loren Feldman, another video blogger with much experience and a loyal following on 1938media.com, explains that video blogging is a valuable communications and marketing method to consider because it is intimate: you can connect directly with your audience and engage with it in an interactive mode unrivaled by text and pictures. He explains, "It's like looking directly at your audience. There's no way you can get closer to your audience than letting them see your face. Nothing tells a story like video."

But how do real videobloggers succeed? What makes a compelling video? The secret to compelling video, Feldman says, is actually knowing who your audience is. In the beginning, you're going to be confined to your own style. You may not get the significant page views initially, just like every starting website. But in due time, you'll learn exactly who watches your content, and you'll know exactly what they're looking for and why they're coming back.

As a video blogger, Feldman encourages people to be comfortable in their roles, and most importantly, to be themselves. The most successful video bloggers are not acting any differently than they normally would. You'll want to connect with what other people are doing and talk to them as if they are your real-life friends. If you're not sure how you look when you do this, start by practicing in front of a mirror. See what your face looks like. Then, practice with a computer and a camera, but don't be overburdened with the minute details of how your camera is situated while you're filming. In fact, keep eye contact with your camera at all the appropriate times.

Feldman also advises against listening to the critics. As your fame grows, you'll likely hear a lot of conflicting opinions, but at the same time, you'll also have that devoted following, so the criticisms shouldn't matter too much. You should not get caught up in the insults, though that may be more of a challenge given that your face is visible. You won't be able to please everyone all the time, nor should you feel that you have to.

The bottom line for videoblogging is that you should have fun. See it as a regular challenge and not as a chore. Realize that this is a way to connect to your constituents nearly face-to-face, and that being personable and up-front is a great way to make that happen.

If you do end up growing a loyal following, you can consider a site like USTREAM (*http:// ustream.tv*) to stream your videos live to the general public. USTREAM is a community where you can share your feelings in real time and follow your visitors' chat simultaneously, and you do not necessarily need to be a videoblogger to do it. USTREAM has been an instrumental tool at conferences and events throughout the world, including the influential keynote speech delivered by Apple Chairman Steve Jobs at the 2008 Worldwide Developer's Conference. You may have heard of USTREAM because of the late 2008 phenomenon of the puppy cam, where six cuddly puppies were raised onscreen in front of millions all around the world.[11] Whether for private correspondence with friends and family or for large-scale broadcasting events, USTREAM is a reliable choice for real-time face-to-face communications.

Evangelizing Content Producers Through Photos and Video

Another way to use video sites and photography is to look at who is already using your product through their creative works and to evangelize these users or encourage that type of activity in the future. Acknowledge their contributions and give them special placement or recognition on your site. In essence, your customers may have already uploaded a testimonial (which doesn't have to be in the written word), so give them the spotlight if they've been passionate enough about your product to photograph or video themselves using it.

In 2007, shortly after the announcement of the launch of Apple's new iPod Touch, 18-year-old budding videographer Nick Haley posted a clever homemade commercial of the new device on YouTube. Instead of threatening legal action, Apple's marketing executive team was impressed. The video was remade in high definition and was broadcast on prime-time commercial spots for several months (*http://www.youtube.com/watch?v=KKQUZPqDZb0*).

If you find that your users are sharing images or video of themselves using your products, show them that you appreciate it. Feature the photographs on your website. Embed the user-created videos on your website. Embrace your customers by acknowledging the content that they create. They invested time in taking that photograph or shooting that video, so let them know that you value their contributions.

Additionally, empower your customers to share their love of your products and services. The makers of Chumby, a consumer electronics product that streams Internet content via WiFi, has a "show off your Chumby" page (*http://www.chumby.com/pages/showoff*) where product owners are encouraged to send photos of their Chumby in the wild (Figure 11-22). The makers of Chumby then highlight these images in dedicated Flickr photostreams.[12]

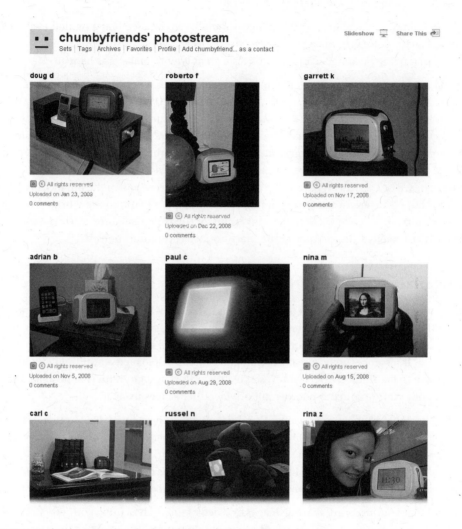

FIGURE 11-22. Chumby engages brand ambassadors via a Flickr photostream

The Emergence of Podcasting and the Podcast Today

Podcasting is a media trend that, while not entirely social (except through audience engagement after its publication), has gained momentum over the past few years. A podcast is an audio or media file that is distributed over the Internet, normally via RSS, for download. Typical podcasters, like videobloggers, offer regular updates of their content for their audiences to consume. Their content is normally provided via audio and they offer downloads, usually via iTunes, for subscribers to then engage and interact.

Videoblogging is not much different than podcasting. However, podcasting is a tad more social, as it makes it much easier to have multiple simultaneous participants on the line. Podcasting is easily done over the phone or via an online service (using a microphone), such as Skype. Podcasting is interactive and social in the sense that most podcasts engage guest speakers, experts, and even the audience. In fact, as podcaster Joe Fowler III puts it, "Podcasting is very social because when you gain listeners, they like to be included, named, or sometimes even invited to participate."

Like any type of online communications medium, podcasting works best when you pay attention to your audience and engage it in your actual program. However, your spoken word may not have the same level of interactivity as the written word, especially since users do not necessarily listen to podcasts while seated at a computer. For example, many avid listeners of podcasts actually download them to MP3 storage devices for listening during a commute to or from work. Further, many individuals shy away from podcasts because audio, in general, is still difficult to absorb entirely unless you can give it your undivided attention.

Therefore, podcasting is a small subset of marketing, but if you're personable (and possibly camera shy) and have lots to say, you may find podcasts a viable extension of your marketing channel. There are some very successful podcasts with hundreds of thousands of followers, though even the smaller podcasts can be valuable and may help further your thought leadership and contribute to heightened brand awareness. If you have the time and dedication to create and maintain a regular podcast, and you feel that your spoken commentary would benefit your audience, you should give podcasting a try.

How Do I Start My Own Podcast?

You can start a podcast for as little as $10 (the price of a microphone for your computer) or for as much as the cost of renting a studio in order to provide a static-free, radio-like experience. Logically, the cheaper alternatives may not necessarily provide the highest audio quality, but all the same, they are good options for starting your first podcast. If you're a success (and if you get sponsors for your podcast), you may be inclined to invest more in better equipment, though it is ideal to start small and cheap before taking the plunge.

One of the best tools for podcasting is Audacity (*http://audacity.sourceforge.net*). This is a free tool that can get you started with recording the stream directly to your computer. Audacity is available for Windows, Mac, and Linux. When using Audacity, double-check that your soundcard actually lets you record what you hear in addition to what you say. This ensures that you record participants in the conversation as well as yourself. Not all soundcards come with this functionality (if in doubt, test before running your podcast).

To engage other users in a podcast, you may want to use Skype (*http://skype.com*) and set up a call. Skype is not an online application—you will need to download it to your computer before operating the tool. Once you have installed Skype, you can initiate phone calls with an unlimited number of participants (though sound quality will degrade with more and more callers). When using Skype, keep in mind that you should always check internal sound settings to make sure that your operating system is controlling the volume and not Skype itself. CallGraph (*http://callgraph.biz*) is a freeware tool that can assist with recording Skype calls. Another method for podcasting is via TalkShoe (*http://www.talkshoe.com*), which you can access over a phone or via a downloadable application.

If you've never podcasted before, take the opportunity before your first launch to record a "pilot" episode. This first recording essentially acts as a dry run. This is a quality assurance method that allows you to listen for technical glitches and hear your podcast before it becomes public. This will help you iron out any issues before the "real" podcast recording, so you will be aware of potential issues and can thus anticipate or avoid them. Podcaster Greg Davies explains, "In a very real way, we recorded the pilot to actively reflect on the flow of the show—and to ensure that we were satisfied with the quality of the content."

When your recording is finished, you may want to clean it up to remove "ums" and awkward pauses. Audacity, fortunately, has these features embedded, so there is no need to search for another audio-editing application.

Now comes the more technical part: you will want to create an XML file to turn your podcast into an RSS file so that you can share it with the rest of the world. Fortunately, a WordPress plug-in called PodPress (*http://www.podpress.org*) can do this for you, without requiring you to know how to create an XML file and all of its semicomplicated syntax. Figure 11-23 shows how PodPress easily integrates into the backend of your WordPress blog.

An alternative to PodPress is Blubrry (*http://www.blubrry.com*), which provides similar functionality to PodPress but is also a community and allows publishers to monetize their podcasts. Blubrry provides statistics for free, though premium features are available for $5 a month.

How Do I Promote My Podcast?

Most podcasts can be promoted via regular social media, though you can publish them on your blog as well with the help of tools such as PodPress and BluBrry. However, the biggest exposure typically comes from iTunes. The publishing process is relatively easy, and with the help of PodPress, the only thing you need to set up is an iTunes account.

Apple has detailed instructions on how to set up your podcast at *http://www.apple.com/itunes/whatson/podcasts/specs.html*.

The promotional tactics we've discussed for other media methods also pertain to podcasting. Leverage your friends, family, and your social media network to give your podcast some momentum. Invite individuals who are experts in the subject matter to participate as well. You can also use social media, such as social bookmarking and social news, to promote the podcast.

FIGURE 11-23. You can embed PodPress into a WordPress backend

It may also help for the podcasters to summarize or even to transcribe the entire podcast, which is useful for those who prefer reading and for the hearing-impaired. Of course, if your personality shines through the audio, your listeners will still tune in when they wish. Since podcasting is one of those communications methods that is is deemed an interruption to the new media spectacle of information overload, you should give your followers something to look forward to—even the ones who prefer to read, because they may still fall in love with your podcast without actually listening to it.

Summary

New media tactics are not necessarily considered true "social media" strategies, but they do have social aspects that can help market your products and services. In the photography realm, image-sharing website Flickr is a great tool for networking, especially if you leverage the service effectively and focus on building community over all other motivations. There are other websites that focus on image sharing, but they do not have as many features as Flickr, nor do they boast the same bustling community with billions of photographs.

If you do not want to put your photographs in the public domain, you can host them on your own website. To do so, it is best to seek out photoblogging software or a gallery application. Keep in mind that your images can be very large, and you may need to purchase additional hosting space if you use this method regularly and store many files on your web server.

On the video front, YouTube is the most popular site, but there are other sites that serve similar purposes. For marketing, you should focus on as many video-sharing sites as possible to spread out your reach. On YouTube, there are numerous ways to promote your video; tags, for example, are important. Your video should not be lengthy, because you're still competing with the rest of the world for viewers' attention. When it's time for promotion, the initial boost is critical for maximum visibility, so don't upload the video and forget about marketing it.

To interact with your audience in a more "face-to-face" fashion, consider videoblogging. Videoblogging effectively lets you "speak" a blog post, and many people find that being themselves is the most efficient use of this communications medium. Similarly, podcasting, a new media marketing technique where you focus more on audio than video, is a great way to engage listeners and build a strong audience.

Endnotes

1. *http://www.smallbusinesssem.com/articles/marketing-on-flickr*
2. *http://news.cnet.com/8301-13577_3-10146825-36.html*
3. *http://www.newteevee.com/2009/01/23/tallying-the-numbers-web-video-rivaled-tv-for -inauguration-views*
4. *http://www.gittinitouttamyhead.blogspot.com/2008/07/fueled-by-fred-kids-want-zipit.html*
5. *http://www.youtube.com/watch?v=QQlzX7EyIwU*
6. *http://www.youtube.com/watch?v=WanLLnVixC4*
7. *http://www.quantcast.com/Youtube.com*
8. *http://www.emarketer.com/Article.aspx?id=1006868*
9. *http://www.comscore.com/Press_Events/Press_Releases/2009/4/Hulu_Breaks_Into_Top_3_Video _Properties*
10. *http://www.alexa.com/data/details/main/youtube.com*
11. *http://www.msnbc.msn.com/id/27724451*
12. *http://www.flickr.com/photos/11410414@N06*

CHAPTER TWELVE

Sealing the Deal: Putting It All Together

Now that you have the basics down for social media marketing, let's focus on putting together everything you've learned so that you can proceed with your promotional strategy for the benefit of the community and for your own marketing motives. We've covered social networks, social bookmarking sites, social news sites, and a variety of new media sites, but for the most part, promoting yourself using only one medium won't yield the biggest success. In fact, for the best ROI, you will likely need to focus your energies on multiple online properties. You may be surprised to see how the community reacts to each different campaign.

Social media is also about declaring your identity early on in the process on the appropriate channels, such as your own website or profile page (but only on services where the community participates in such activity). It is also about creating a communications channel that you actively monitor and participate in. Communications in social media do not necessarily have to happen onsite; the real-life relationships that follow can benefit your company objectives, so use social media as a stepping stone to find other means of communication.

Identification: Telling People Who You Are

The first step in social media marketing is to be open and honest about your reasons for participating in the space. Who are you and why are you here? Typically, this will be something you declare on your social networking profiles (where you represent yourself as an employee of a particular company) but also on your website's About Us page. Use this page to articulate your company's mission statement or to express your role as an employee of the company. If your organization has many individuals who are all active in this big Internet space, share their bios on a single searchable page so that other people can learn more about them.

Once you have declared your identity on this space and have become ingrained in the social media sphere, let people know where to find you offsite. For example, use your biography to share your social media profiles with your readers. Let them know where to find your company and its employees on LinkedIn, Twitter, YouTube, StumbleUpon, delicious.com, Flickr, and all other social media sites. This way, your constituents can affiliate themselves with your brand on yet another level. You may already see an intersection of traditional business and social media, as Twitter is now becoming "mandatory" on business cards,[1] along with other information about you (Figure 12-1).

FIGURE 12-1. Not everyone limits their business cards to just their Twitter usernames!

The Social Media Workflow

In 2007[2] and 2008,[3] UK agency Immediate Future performed studied the involvement of big brands in social media. The studies, referenced in Chapter 2, show the impact of social media on these brands in terms of online interactions, sentiment, and subsequent visibility. Immediate Future looked at some very big brands that have fared well with social media involvement, including Google, eBay, Canon, Porsche, Intel, MTV, BMW, and others, and measured how active they were in a number of social media channels across the Internet.

According to Immediate Future's information (shown in Figure 12-2), not one company focused entirely on one single avenue to promote its business objectives. Instead, the brands spread out their involvement across different channels. This makes sense, as the appropriate goal of any social media marketing campaign should be to reach different demographics and users. After all, how many people do you know who actually engage regularly across Digg, YouTube, Flickr, MySpace, delicious.com, Twitter, and Bebo (which is popular in the UK where this study was performed), in addition to regular blogs and forums? Chances are, you don't see that many people who are addicted to all channels at once. I know a few "social media experts," but even those individuals are not well versed in all the nuances of all social networks and do not participate on a consistent basis across all of the sites.

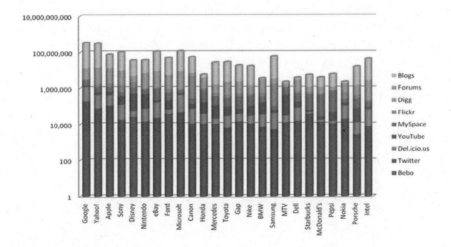

FIGURE 12-2. Brand share of voice in detail

Therefore, it's important to think beyond a single social media channel. If the big brands are doing it, it's for good reason. The reach of each individual sites only extends so far; by touching upon additional channels on the Internet, you're able to extend that reach even farther.

Revisiting Return on Investment

In Chapter 1, you learned that it can be difficult to get actual numbers for determining ROI on your social media campaign; you can't put a numeric value on the buzz and quality of an online conversation. However, there are other ways to measure the success of your social media marketing efforts. We've reviewed a number of tools that can show you if your campaign is successful. Note, though, that it is not easy to directly correlate conversions with your social media marketing strategy, especially if you have several marketing strategies in play simultaneously.

There are five separate metrics that you can look at to estimate your ROI:

- Reach
- Frequency and traffic
- Influence
- Conversions and transactions
- Sustainability

Reach

How far is your message traveling? You can determine this by the number of links your story has garnered, the number of people tweeting about your campaign, or the number of connections you've accumulated since you listed your Fan page on Facebook. Depending on the channel, you can measure this by seeing how many retweets a specific story or URL has gotten, or you can review a particular URL by using Yahoo! Site Explorer (*https://siteexplorer .search.yahoo.com*) (shown in Figure 12-3) and noting how many "Inlinks" (Yahoo! terminology for inbound links) are now pointing to the URL in question.

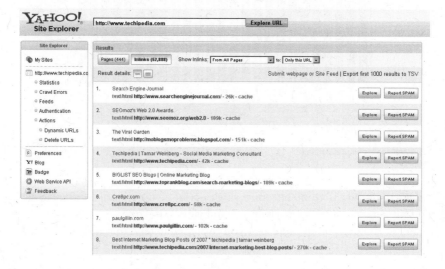

FIGURE 12-3. Yahoo! Site Explorer counts and shows inbound links to a particular URL

Frequency and Traffic

How often are people visiting your site? To determine this, open up your analytics software and look at the number of impressions in a specific time period versus other periods. If you see a surge of traffic and you are doing no other marketing at the same time, it may very well be attributed to your social media marketing campaign. You may also want to review your web analytics to see how many visits your website receives on average and how frequently you received visitors after your campaign took off.

Influence

How deep are conversations related to your business? Are people actually discussing the subject, or are they looking, commenting (or not), and moving on? If there's more depth and influence, there's more potential for conversion and virality.

Conversions and Transactions

Are you actually seeing people click through to other parts of your site since you launched that viral piece to raise awareness about your business? Are they downloading the software you've asked them to try? Are you seeing additional transactions? Registrations? Purchases?

Sustainability

How long will users stick with you after your social media campaign gets plastered on their radars? Are they going to stick with you once they become aware of your existence, or are they going to go elsewhere? Will you have them for a short while until their involvement tapers off; will you see them only for the duration of the campaign; or are they true customers for life? Figure 12-4 shows what you can expect of a successful viral launch, though results may vary. Before the article shown took off, there was not much traffic generated on the site. After the article was passed along, the site saw more traffic.

You can gauge the success of a social media marketing campaign by looking at these different metrics. However, your internal marketing team should review each individual metric and determine how to best measure the output, so there is no definitive suggestion about what to look for besides results. Plus, in alignment with your SMART goals (see Chapter 2), you may need to tweak metrics.

While some of these areas may seem a little broad, you can often assess the overall success of a campaign by reviewing the quality of the reach and influence. For example, influence is something you may not be able to measure directly, but what is the quality of conversation? Are people considering actually trying out the product after being alerted to its existence? Is the promotional message causing users to shy away from your product? The idea here is to start *listening* to the conversation, and then participating to create long-term relationships that will yield success in the future.

Am I Done Yet?

Your big social media marketing viral campaign may have taken months to formulate and execute. You typically won't want to stop there. Regular interaction is vital. Consider this logic:

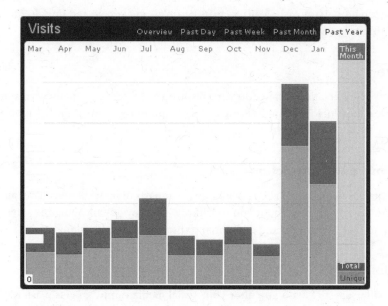

FIGURE 12-4. Web analytics: Traffic before and after a short-term viral campaign

do you write a blog post and just call it a day after that post has gained momentum? If you don't write consistently, your subscribers will stop coming to you for more content. According to a survey conducted by blogger Darren Rowse,[4] 29% of respondents said that the biggest reason they unsubscribe from blogs is because the blogger does not update his or her blog frequently enough. If you don't keep your information fresh, many readers will move on to bigger and better things.

As you can tell, you have a lot of work to do, but it gets easier as you gain credibility in the space. Plus, like real-life face-to-face relationships, online interactions eventually become something you can do without effort as long as you practice at it at first. You may already be a pro without knowing. It doesn't hurt to take the plunge.

Committing Yourself for the Long Term

A one-time deal may work for that two-week promotion, and if that is your goal, you should stop right there. For most individuals and organizations, however, a long-term engagement will require a long-term commitment. Just like any type of promotional tactic, social media marketing is hard work. You can work on viral campaigns, but after that, you will also need to be entrenched in the community and involved in the discussion. You may not want to focus on viral campaigns at all, but if you're considering social media, you'll still need to be social. Thus, involvement in the community is important.

Never Stop Listening

Listening to the conversation is something you will need to do on a regular basis. Staying silent is a good way to shun your followers and cause them to turn to another provider for the same (or similar) services. Don't be one of those social media marketers who engage only when it's critical to silence the naysayers; have a consistent dialogue.

Whether or not you want to participate, you should always be monitoring for mentions of your brand, CEO, public relations people, and industry, and you should use this information as a way to enter the conversation. You can build your brand simply by listening and subsequently participating. Add meaningful comments to blog posts. Involve yourself in the conversations on Twitter, reddit, Facebook, and FriendFeed, especially as they relate to your company. Pay attention to what people say in your industry and use this information to grow and better yourself and your offerings. On that note, what are your competitors doing? Are they listening? Are they not? If they are, can you do it better? If they're not, can you be the first person to start the trend?

Consistently review and monitor mentions of your company. Watch the sentiment. Is it positive? Is it negative? Can you turn that sentiment around? Try to nurture the opinions of people around you.

Remember the Community

All creators of social media sites had one thing in mind: they wanted to connect people with similar interests or to make it easier for friends, new and old, to communicate with one another. At their core, these social networks still exist for this purpose. You must take this into account at all times when considering marketing in these communities.

If you've worked in an advertising or public relations agency for several years, your mindset may be, "I'm looking to market my product, so let me get people's email addresses and start sending them newsletters." Well, that might be your strategy if you aren't aware of how to embark on this different type of journey into a world where people who accept your friendship are not necessarily opting in to communications (plus, this is a violation of the terms of service on multiple social networks—read the fine print!) Ask a 20-year-old Facebook addict how you should market your product in her territory (remember, Facebook started off as a social network for college kids), and her answer may be much different than you expect. Overt marketing tactics will not be well received by the youth and active social media users of today. These old strategies may end up causing people to push you away. This, in turn, could also cost you your reputation, and in a world where reputation management failures can really break that business you've spent years and generations making, you don't want to take that risk.

Never look at a social media site solely as a means to promote your product. If that is your goal in social media marketing, you may find that your campaign is a failure. Bring someone on board (that 20-year-old Facebook addict, perhaps) who cares about the community she is involved in. These enthusiasts and avid users are good resources to tap into; nobody already established in these communities would want to ruin their credibility by involving themselves in inauthentic interactions. Only those genuine relationships prevail, and these individuals already have the following and credibility to make your campaign a success. The bottom line is that selfish motives won't find success in social media marketing, and the companies and individuals who only take and do not give back to the community will be weeded out.

Jay Izso, a 40-something in social media, shared a story with me about a time he was befriended by an individual on LinkedIn, and he immediately accepted. The next day, that person started promoting his business and services to Izso. Apparently, this new "friend" was only looking for people to market to. What did Izso think? Well, he shared that story with me—it obviously left a bitter taste in his mouth. If you're going to join social networks primarily to evangelize your business, you're going to prompt people like Izso to share your etiquette missteps with others. Be real first. Let people become curious about what you have to offer, and then you can start marketing to them.

The bottom line is that you can achieve success in social media marketing, even if you're still focused on a traditional marketing mindset. You can learn by doing and by watching how others participate. Do what other successful community participants are doing. Rinse and repeat. Keep in mind that success takes time. You must be willing to devote that time to the task.

Strategizing in Social Media Communities

Social media includes blogs and online sites (video, social news, bookmarking, etc.). By now it should be evident that you need to make your presence known in multiple areas.

Your Blog Is Your Hub of Communication

Consider your blog as your home base. This is where you can freely communicate your own objectives and take the community's questions for future consideration and deliberation. You may receive feedback, so make yourself accessible by contact form or via comments.

Build Your Brand on Other Social Properties

Build your company's presence on online properties beyond your blog. Start that YouTube video channel to upload videos. (If you're not up to snuff on video production, at least grab a username for your company while you can. Check availability of usernames on social sites using a site like *http://www.knowem.com*; Figure 12-5). Create your free Flickr account to share your photographs, and when you find that it's helping you and you're getting lots of visitors and have more photos to share, consider upgrading to a pro account. Start registering your username on other social media sites. Don't use your company name on a site like Digg

FIGURE 12-5. Claim your name for social media services via KnowEm, even if you don't plan to be active on all of them

or reddit, though, especially if you plan to submit stories on your company's behalf there. If you do, it will smell like marketing and the community members won't have it.

NING: CREATE YOUR OWN SOCIAL MEDIA COMMUNITY ON THE CHEAP

Ning (*http://www.ning.com*) is a different beast from the communities discussed in the previous chapters. This free service lets you actually create your own social network that includes photos, a forum, videos, and more. Community members can engage with one another via messaging and real-time chat, and as with Facebook, network creators (group administrators) can add utilities and games to the individual network using third-party OpenSocial applications (accessible via *http://www .developer.ning.com/opensocial/application/list*).

Ning is successful because it serves a need: people congregate around specific topical interests. As such, there are over one million topical social networks on the platform. This provides a great opportunity to create a network around your product or industry (if one does not exist already) and to invite members to participate. Ning sends more than six million unique visitors per month to its niche sites, and this fact proves that people actually want to discuss matters that are important to them. It also illustrates that human capital is cheap; you don't actually have to invest in a full-fledged enterprise solution that becomes a social network.

Keep Your Options Diverse: Don't Fixate on One Community

Putting all your eggs in one basket is never an ideal strategy. While it may be easier to become an expert in one site, you should try to spread yourself across multiple social media sites (but don't spread yourself *too* thin). If you find social media an imperative strategy, allocate more than one individual to the task. Use someone who is well versed in photography and video production. Use someone else who is well versed in copywriting for social media, and use yet another person who is well versed in social media promotion across social bookmarking and social news sites. On the other hand, consolidating all of these roles among one or two people may be the only way you'll see results, especially in small businesses where manpower isn't abundant. Either way, bear in mind that it's a time commitment that you must be willing to invest in at all times.

Adopt a Social Media Mindset

Start adopting a social media mindset throughout your organization. The CEO of your company should understand social media enough to involve himself if necessary, though heavier involvement would be well received by constituents across those services. Even maintaining a personal profile on LinkedIn and a Facebook profile page is a good way to start. Have him contribute to the blog, time permitting, and perhaps he can speak via video about the innovations the company is making for both employees (for your internal newsletter) or for the actual users of the service.

Build Those Strong Relationships

To promote your organizational goals, build genuine relationships with users of social media communities. Share your interests with them and let them get to know you. Be real and genuine, and long-term relationships that go well beyond social networks are likely to follow.

Thinking Outside Social Media Communities: Face-to-Face Interactions

You have already established yourself as a credible member in the community and have created an online identity for your constituents. Now, take that one step further: if you've already started initiating relationships online, how about taking those relationships offline? Graco Baby (see Chapter 4) did this by inviting parents and children to participate in offline family-friendly events it created. Mashable (*http://www.mashable.com*), a blog covering the general Internet and social media space, regularly hosts events in both San Francisco and New York City; the blog also accommodates readers outside its two central locations by traveling to other cities. Readers of the blog are invited to mingle and interact and often are presented with demos from other companies in the space. Whether as sponsors or attendees, companies can take advantage of these actual events to network with new prospects, especially if they do not have the budget to host large-scale events themselves.

In reality, nothing is better to solidify a relationship that has evolved online than to network face-to-face. You'll also find out that you have a lot to talk about, from collaborating to sharing feedback to providing valuable insights and information. Many of my colleagues consider themselves to be successful because they've gone out and actually networked with prospects. Have you?

If you are interested in taking advantage of real-life interactions, consider searching for local events in your area that might be a good fit for you. Use sites like Upcoming.org (Figure 12-6) to find out about events in your area or elsewhere.

FIGURE 12-6. Upcoming.org lists events based on certain search criteria

Additionally, you can join a group of like-minded individuals via Meetup.com, a site that unites individuals with similar interests through real-life, face-to-face interactions. Figure 12-7 shows that you can join 13 groups related to "diabetes" and then find out about local events in your area.

At these events, network with people who are most interested in the topic, and you'll be able to establish thought leadership (and brand awareness). These real-life interactions are a great way to meet your customers and find others interested in your product offerings.

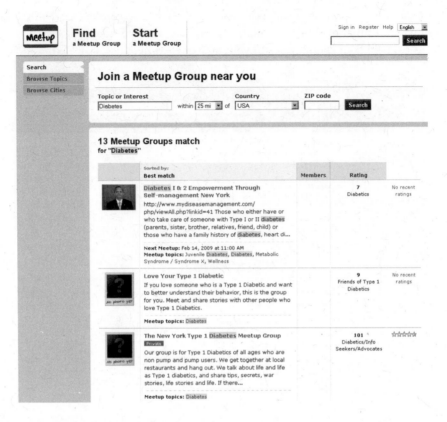

FIGURE 12-7. A list of groups hosting events through Meetup.com

Fostering Creativity Online

Still, social media is all about being social online. How do you actually find something that will work in the social media marketplace?

The key to success is to think outside the box. Do something you may not expect others to do. This strategy worked effectively for UK-based skin cancer charity Skcin, which launched the Computer Tan website (*http://www.computertan.com*) in early February 2009 (Figure 12-8). The premise behind Computer Tan was that individuals could get tanned through the rays emitted by their computer monitors.

The Computer Tan website was a hoax. However, over 30,000 individuals in the UK signed up to participate in the program within 24 hours of its launch.[5] At the end of the day, these users were informed of the dangers of tanning and the harmfulness of the sun's rays.

While this tactic may be one of the shadier strategies involved in viral marketing (and I'm not sure the 30,000 individuals who fell victim to this hoax were appreciative of its true nature),

FIGURE 12-8. Computer Tan's home page: unexpected viral innovation

this is the exact type of brainstorming you can use to create a memorable and successful social media strategy.

If great ideas don't come to mind, consider the following successful strategies from social news and bookmarking sites.

Viral Strategy #1: Lists

1. This list is going to be awesome.
2. I'm going to tell you why shortly.
3. You're going to love it.

Read the preceding sentences. They're broken up for easy reading. If I provide the same sort of information in paragraph form, you may not be able to actually absorb each individual segment. In fact, lists (Figure 12-9) often perform better than paragraph posts or articles.

Lists are viral by nature because they encourage heavy engagement, conversation, and communication, and they also often show that the writer has done some extensive research on a specific subject. With a list, you can comment on a single item listed; it's a lot easier to isolate and nit-pick those facts you may not necessarily agree with (or, on the other hand, that you agree with a lot). By their nature, lists are:

- Scannable and thus easily digestible.

- Typically short and the content therein is easy to consume. (If you write a long list, emphasize the main topic of each bullet point in bold before going into specifics.)

- Resourceful and can provide a great deal of information in a single article. Lists, therefore, can serve as references for a later date.

- Meant to be shared, thereby increasing traffic and links, and can help increase awareness.

- Engaging, and encourage individuals to participate.

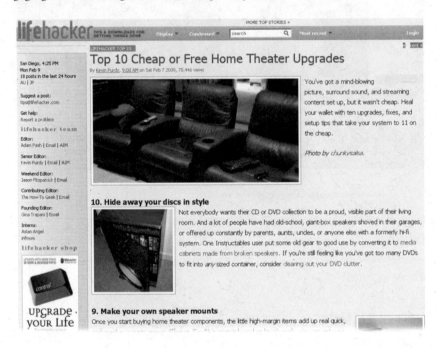

FIGURE 12-9. Lists succeed in social media because they're heavily referenced

The catch to lists is that they are a bit overused in social media channels, so don't turn every single article or post into list format. Similarly, don't break up your list into several pages if you can avoid it. On some online publications, the extra page views help boost advertising rates, but if you're a blogger or writer who does not utilize such advertising, it's best to steer clear of that kind of list implementation. Most readers may want to either print out the article or to reference the content easily, and adding additional page views that obstruct and complicate their access to the article can cause frustration.

Viral Strategy #2: The Quiz or Questionnaire

Are you obligated to share only articles in social media? Not necessarily. There's video, photography, and even short 140-character blurbs that take seconds to digest. Additionally, there's the quiz or questionnaire. Give people a chance to participate by answering questions about themselves on a variety of topics, from "Are you a romantic?" to "Are you addicted to the Internet?" to "How bad are your spending habits?", and you've got potential to share something about a specific subject that can ultimately go viral.

You can make up interesting questions that are related to your business. Then, give people the option to participate via true or false answers, by multiple choice, or by providing their own input (Figure 12-10).

FIGURE 12-10. The quiz: give people the opportunity to share, and score them based on their answers

Once the quiz or questionnaire is completed, give people the option to share their results with their peers. They could simply share the data with their friends by linking directly to the quiz. On the other hand, there's a lot more potential for the quiz to be spread if you create a widget that quiz participants can place on their sites via HTML code once they've completed it, as shown in Figure 12-11. You can also extend quizzes to social networks like Facebook; quizzes

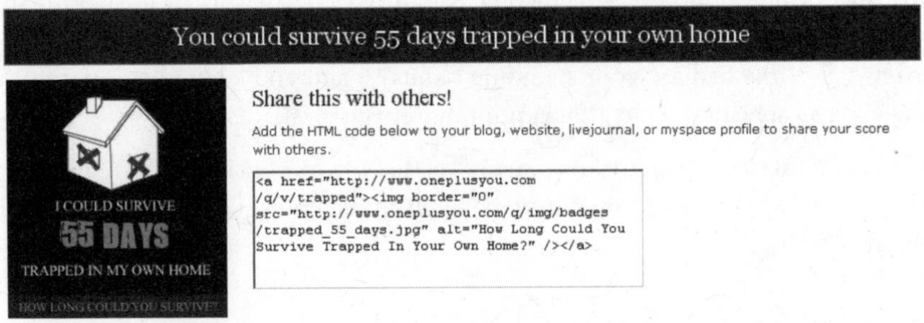

You could survive 55 days trapped in your own home

Share this with others!

Add the HTML code below to your blog, website, livejournal, or myspace profile to share your score with others.

```
<a href="http://www.oneplusyou.com
/q/v/trapped"><img border="0"
src="http://www.oneplusyou.com/q/img/badges
/trapped_55_days.jpg" alt="How Long Could You
Survive Trapped In Your Own Home?" /></a>
```

FIGURE 12-11. Give your quiz participants the ability to share their results on their websites, social media profiles, or blogs

are often popular on that service. The best part about these HTML codes is that, by nature of their existence, they help build relevant links to your site, especially if the quiz is related to your business objectives.

Viral Strategy #3: Participatory and Interactive Video or Games

If you can get a graphic designer to create an interactive game or video that promotes your product, you should definitely pursue that opportunity (see Figure 12-12). Let people personalize pages with your product to share with their friends.

FIGURE 12-12. An interactive holiday card

Interactive video that users can personalize may be the costliest option, but the return is substantial: users find that they can really engage with your brand while also being able to provide information about themselves. They are compelled to share these videos and games with their friends, and the result is extremely powerful word-of-mouth marketing that works by virtue of social interactions.

Some examples of successful participatory and interactive video include the Gillette ManQuarium (*http://www.manquarium.com*) (shown in Figure 12-13) and in the promotion of Bob Dylan's album titled *His Greatest Songs*. In the former, participants search an online database for the "perfect body" for their "perfect-looking guy" and plaster their crush's face onto the body. Once the guy is created, participants are asked questions about their crush and can then interact with their guy. The end result offers increased branding for Gillette as well as entertainment for the participant.

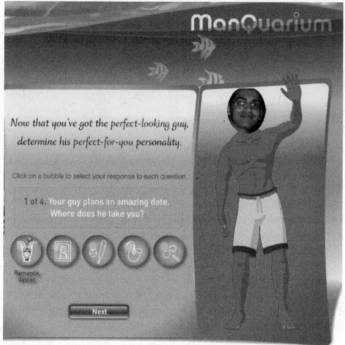

FIGURE 12-13. The Gillette ManQuarium interactive game

In an effort to promote Bob Dylan's new album, one organization created an interactive game in which users could fill out 10 cards; a video would then launch depicting Dylan flipping through the customized cards to the tune of one of his songs. Figure 12-14 illustrates the game.

Once the cards were filled out and the music video started playing, users could send the video to their friends, which meant that as more and more people received it, there was more and more potential for it to be passed around. This is word-of-mouth marketing at its finest.

FIGURE 12-14. Bob Dylan's interactive messaging game

Viral Strategy #4: Tell a Story with Images

As children, we grew up preferring illustrations in our printed books. As adults, we still have an emotional attachment to strong visuals and compelling imagery. In an online society where there's an excess of information, mostly in the written word, some individuals prefer content that is dressed up with images, as an image can contribute to a story's success. In a social marketing atmosphere, visuals and images tell powerful stories.

If, for example, you work for a data recovery firm, wouldn't it be interesting if you shared photographs of the hard drives submitted by some of your customers? You've likely had customers who have had their computers burned in fires or smashed by exes, so why not share the customer's story and use the image of the hard drive to show your readers exactly what became of their precious data? On the other hand, what if you work in a drug rehabilitation home? You may want to bring awareness of the dangers of drug dependency by chronicling the last days of a heroin addict's life[6] in a photo journal of very emotionally provocative images.

The example image essay in Figure 12-15 shows the photographic journey of a mother celebrating the final days of her son's life as he battled cancer. The story itself is gut-wrenching, and some of the related photographs may even make you cry. However, the mother's affection toward her ailing son was clear, and the love of her little boy was shared among thousands. The essay won the Pulitzer Prize in 2007 for feature photography.[7]

FIGURE 12-15. A photographic journey of a mother's last days with her son as he battles cancer and succumbs to it

These examples are testaments to the power of using images to raise awareness of different subject matters, but providing commentary, too, can bring additional success to your viral marketing initiatives.

In a similar fashion, if your customers are interested in learning more about the process of manufacturing your products, you may want to give them a tour of your factory or office. Your followers and stakeholders would be most interested in visualizing the company dynamic, which becomes a great opportunity to share what goes on behind closed doors in a video or photo tour.

Viral Strategy #5: Build a Tool

Does your industry suffer from a problem that you may be able to solve? Perhaps that need is not yet on the top of your mind, but chances are, there's something you can do that will help make others' lives (and even your own) easier. Building tools is one such example. In fact, tools are a terrific way to build high-quality relevant links to your site and to establish thought leadership. For example, do you specialize in computer infrastructure for enterprise management? Perhaps you want to provide a tool that will help webmasters monitor their servers, like the one from Dotcom-Monitor[8] shown in Figure 12-16.

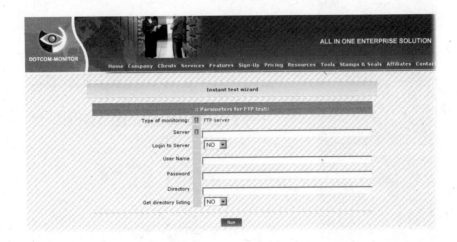

FIGURE 12-16. An online tool that monitors websites

Think about the big online tools that already exist: you've probably seen a mortgage calculator or a calorie counter on a website or through a search. If your industry addresses either of these categories, can you use the same tool's concept and make it even better? Can you think of another problem along those lines that your users may want to use the Net for to solve easily? These applications benefit just about everyone, even those not necessarily looking to buy a product, and they have the added benefit of being shareable and thus helping promote brand awareness.

The iPhone, Palm Pre, and other handhelds allow for the installation of third-party web applications, and these applications are more ubiquitous than ever. You may want to empower mobile-dependent users by giving them a solution that they can access at any time and from any place. Mobile utilities are becoming a lot more prevalent, as they can help users solve everyday problems. In addition, individuals can easily find mobile applications outside your industry's website (such as through iTunes, for example), and you may accumulate new interested parties without overtly marketing to them. In essence, your new followers will be able to find you easily without even actively searching for you.

Viral Strategy #6: Teach Your Users How to Do Something

Do people look to you for advice? Teach your users how to do something. The best way to do so is to utilize video for the purpose of illustrating a process. For example, there are a number of great videos out there that teach people how to tie a tie,[9] how to fold a shirt in less than 2 seconds[10] (shown in Figure 12-17), and how to shuffle poker chips like a professional.[11]

YouTube — Broadcast Yourself ™
Worldwide | English   (0) ✉

| Home | Videos | Channels | Community |

Japanese way of folding T-shirts!

左手でつまんだ所から延長した
肩と裾を右手でつまむ

0:07 / 0:38

Rate: ★★★★★ 658 ratings   Views: 762,668

Share   ♥ Favorite   Playlists   ⚑ Flag

✉ Send Video   MySpace   Facebook   more share options

FIGURE 12-17. A video showing how to fold a t-shirt

Videos can really work to promote thought leadership. Further, the most compelling of the videos may be referenced on how-to blogs and on other sites.

In Chapter 4, you learned about the social media methods utilized by the Caminito Argentinean Steakhouse. Its YouTube channel is filled with informational food videos teaching budding chefs how to prepare great meals. For example, it offers videos titled "How to Make Fire-Roasted Sweet Onions" and "How to Make Mate."

There are infinite possibilities when teaching someone how to do something. You can illustrate the entire process through a series of photographs and explanatory captions, or you can use videos to speak directly to your readers (in a method similar to videoblogging) to teach them how they can do something well.

Encourage "Old-School" Tactics, Too

While social media is about online collaboration and sharing via newer technologies and methodologies, do not forget about those traditional tactics that gave rise to the social media marketing of today. Old technologies are becoming increasingly "interactive" in the sense that they encourage others to participate in these new marketing initiatives.

- As mentioned earlier in this chapter, your business card is a great way to self-promote and share your blog, website, Twitter ID, and all social media profiles.

> **TIP**
>
> If you promote your social profiles in your business card, you want to pursue the mindset "less is more." Think about emphasizing only the social profiles you are active in and want to call attention to. You don't want to overload the recipient of the card with too much information. As a compromise, if you maintain social media accounts on multiple services, highlight only the strongest and most active.

- Send relevant marketing initiatives directly to your peers through email or Instant Message. On the same note, if you have a newsletter, utilize it to promote these stories, videos, pictures, or other related viral campaigns. Ensure that the recipients of this messaging have opted in; you do not want to overwhelm unsuspecting individuals, as this will tax the relationships you currently maintain.

- Use the "send to email" features on social websites. This a little-used tactic that still performs effectively, though you should use it sparingly to avoid overwhelming your friends and family with too many "look at this!" requests (remember, your not-so-social friends are also dealing with the challenges of information overload).

- Your email signature is a great way to promote your social media initiatives as well. You shouldn't go overboard by adding every single social media profile, but you can highlight the active profiles. My own email signature is as follows:

  > Tamar Weinberg
  > *http://www.techipedia.com*
  > Twitter: *http://twitter.com/tamar*
  > FriendFeed: *http://friendfeed.com/tamar*
  > This email is: [ ] bloggable [x] ask first [ ] private

- Use forums. Forums themselves are social, but they predate the social media of today. Still, forums follow the same rules of social media marketing; you will need to be absolutely genuine and involved in the conversation to be able to promote without running into problems with the community. You will likely also want to be somewhat established in the community before immediately promoting something so that your posts are not perceived as self-serving.

By also focusing on more traditional avenues outside of the social media world, you may receive additional eyeballs on your marketing efforts. These people may not be active in the social networks that this book discusses, and may otherwise not know about the content promoted on these other channels.

The Bottom Line

When it comes to social media marketing, the rules of engagement are different. Altruism rules above all. Authentic online relationships can further your cause and help foster real relationships that can flourish offline. In fact, you need to always think about the relationships before you think about the marketing goals, which in most cases will be your bottom line for participation on these services (despite the fact that they are just fun in and of themselves). Your first step should be to seek neighborhoods of shared interest. After you determine that there is curiosity in who you are and what you do, that's the appropriate time to seek a business relationship. That curiosity may not always come, but you can always participate by overtly identifying yourself (where appropriate) as someone who represents a certain organization, company, or industry.

If you plan on promoting yourself and your company, don't forget to consider the people around you. Acknowledge them by listening and responding. If you're a blogger, link out to them and highlight their positive and valuable contributions to your blog, especially if you were particularly moved by their commentaries. They took the time to read and respond, and acknowledging them is a great way to show them that you appreciate their contributions. Showing that you care is a big part of being social, and helps tremendously in social media marketing and fostering relationships, which ultimately can translate to powerful marketing. In fact, the best way to seek out promotion is to promote others before you promote yourself.

Online interactions are different than they were in the past, and being personable is the key to succeeding in this day and age. For example, the standard press release does not work as effectively as an email message where you personalize the message and show the readers that their attention really matters. Public relations executive Todd Defren exemplified this strategy with one of his clients. In an effort to promote a client's product, Defren's team took the time to write very personalized messages to mainstream bloggers.[12] Defren's group often tied the client's messaging into a previous post on each individual blog, and it was clear that each pitch was well researched and written with the specific blogger in mind. In the end, the messaging was very well received and the client's product enjoyed a lot of mainstream press.

Defren's example is a brilliant one, but is one that many public relations professionals do not take time to follow. The problem with many PR agencies is that they only have their clients in mind. Bloggers, however, have to wade through hundreds of pitches on a daily basis and are overwhelmed with incoming press releases. The pitches that succeed are the ones in which the blogger or blog is put first. Marketing motivational speaker and writer Seth Godin summarizes the personalized approach quite nicely: "If you have more than a few people to contact, you'll be tempted to copy and paste or mail merge. Don't. You'll get caught. It shows. If it's important enough for someone to read, it's important enough for you to rewrite."[13]

Social media marketing is not an "easy way out." Sweat and hard work is required for the ultimate success. Whether that means researching a community thoroughly to determine the exact messaging, networking with people who can influence the success of your marketing message, or writing thorough blog posts or emails that really engage the reader and community, this is not something that you can simply do in 12 hours' time. You must be willing and able to commit time to the task and consistently work toward building your brand in the eyes of your beholders.

This is the new era, and this is now.

Summary

Social media marketing is a comprehensive effort that requires interactions across multiple online properties. When you engage, be open about who you are when and where appropriate. It's also important to acknowledge that social media is about genuine conversation and communication, and while these are tools to help you achieve that goal, social media marketing goes beyond just utilizing these tools—it is about empowering the voices of both the producers and the consumers. As such, there are many ways to achieve these goals. Tools act as facilitators to make that happen, but the proper mindset should be there all along.

As you have seen in this chapter and throughout the book, ROI is not necessarily easily traced back to social media, but you can focus on several different metrics to estimate success. These are reach, frequency and traffic, influence, conversations and transactions, and sustainability. However, even after you wrap up a social media marketing campaign, consider monitoring via listening and responding. Don't wait until there's a crisis to get involved. Focus on keeping your content fresh and current so that people will have reason to interact with you. Even if you don't have a new product release on the horizon, you still have something of value to share with your constituents.

The community should always come first, a premise that should be understood and followed throughout your organization. Those who promote but do not understand the feelings of those around them will not see their messages travel very far. If this concept is difficult to understand, it may be useful to hire someone who eats, breathes, and lives in the world of social media, especially in the networks in which you are attempting to market. Those who really are involved in social media sites will often find a way to self-promote (or promote your business) without being perceived as selfish, because they're likely giving back to the community as well. If you cannot hire someone who can teach you the path, learn by doing, but be sure to follow the examples of others who have been successful.

Online communities exist in abundance, but there are four main types of social sites (social networks, social news sites, social bookmarking sites, and new media/video/photography sites). Don't limit yourself to one category; be everywhere. While you don't want to spread yourself too thin, you should definitely maintain accounts on different types of sites and have a general understanding of how they work in the event that you need to share something worthwhile.

Networking online is the majority of the social media marketing battle, but taking it offline is even more powerful. In a face-to-face relationship, you can have real interactions that are genuine, forthright, and can truly make a tremendous difference. Don't shy away from asking questions and soliciting feedback or learning about other people. Have an open mind and be willing to learn. If you're ready to take the plunge into the real world, use sites like Upcoming.org and Meetup.com to find events in your area that are relevant to your industry. Attend conferences and trade shows on these subjects. Never be afraid to network—it's one of the most powerful tools for marketing, period.

If you're aiming to raise eyebrows via a computer monitor, start thinking along the lines of viral marketing strategies, which, by nature, can be shared and redistributed. List articles and posts, for example, are successful because they encourage participation and sharing. Quizzes let people provide information about themselves to you and to their friends. This often prompts others to participate, and the result is that people are often paying it forward as they pass the meme around. Similarly, interactive games allow users to identify with specific brands, and in the case of distributable personalized games to share their creations with the world.

Furthermore, a really compelling photo journal can evoke heavy sentiment and emotion. Some of the most powerful viral photography marketing pieces chronicle the lives and deaths of individuals. Other enlightening photography tours can walk viewers through building or manufacturing processes. In photography, sometimes the images speak for themselves, but commentary can also help.

Tools are yet another way to make it easier for someone to do something, so if you can solve a problem or fulfill a need by creating an easy tool, by all means, go for it. You can also make people's lives easier by using a video demonstration to show them how to do something. If you can do it, why not teach others? This strategy definitely helps establish thought leadership.

The aforementioned tactics can succeed for social media marketing among the various sites discussed throughout this book, but don't overlook the old tried-and-true tactics of sharing via traditional means, including forums. Some adherents consider forums to be social media, even though they are far older than the social networks and social sites of today. Other traditional strategies include using email signatures (or simply sending out email messages) and IM. Of course, if you've looked at a city bus lately, you'll likely see URLs all over the place; similarly, your social media URLs do not necessarily need to reside on the Internet. You can plaster them on business cards and throughout your print promotional materials.

The bottom line is that social media marketing is about real, genuine relationships. Give of yourself, and others will give back to you because they value what you do. Show others that you appreciate that they are listening to your messaging by putting them first. Also, remember that social media is not a "get rich quick" scheme. Like any marketing channel, it takes time and effort to yield the best results in this space. Your success will come from continued effort and dedication to the task.

Two-way conversation is here, and it is now. Talk to people, listen to them, and keep in mind that it's not much different than having a real, face-to-face conversation. You truly do not need to fear the big world ahead of you.

Endnotes

1. *http://www.daggle.com/090204-090732.html*

2. *http://www.immediatefuture.co.uk/the-top-100-brands-in-social-media*

3. *http://www.immediatefuture.co.uk/the-top-brands-in-social-media-report-2008*

4. *http://www.problogger.net/archives/2007/03/01/34-reasons-why-readers-unsubscribe-from-your -blog*

5. *http://www.technology.timesonline.co.uk/tol/news/tech_and_web/article5667995.ece*

6. *http://www.fishki.net/comment.php?id=26844*

7. *http://www.pulitzer.org/works/2007,Feature+Photography*

8. *http://www.dotcom-monitor.com/task_instant_test.aspx*

9. *http://www.youtube.com/watch?v=MbXzI-IAdSc*

10. *http://www.youtube.com/watch?v=b5AWQ5aBjgE*

11. *http://www.5min.com/Video/Shuffle-Poker-Chips-Like-a-Pro-8284*

12. *http://www.pr-squared.com/2009/02/blogger_relations_and_social_m.html*

13. *http://sethgodin.typepad.com/seths_blog/2009/01/how-to-send-a-p.html*

The Ultimate Social Media Etiquette Handbook

I originally posted the "Ultimate Social Media Etiquette Handboook" on my blog at techipedia.com[1] in December 2008. It became so popular that I have included it here as an appendix.

Social media mimics real relationships in many cases. Would you do the following in real, face-to-face relationships?

- Jump on the friendship bandwagon without properly introducing yourself?
- Consistently talk about yourself and promote only yourself without regard for those around you?
- Randomly approach a friend you barely talk to simply to ask for favors—repeatedly?
- Introduce yourself to another person as "Pink House Gardening?"

If you answered "yes" to any of these questions, you may need a refresher course on social media etiquette—and perhaps real-life etiquette also. Here are some egregious sins that you must not perform on social media sites. Avoid these violations and learn how to manage and maintain online relationships on a variety of popular social media sites.

Facebook

- Adding users as friends without proper introductions. If you're looking to make friends, tell people who you are. Don't assume they know you, especially if they don't.

- Abusing application invites and consistently inviting friends to participate in vampire games. Many call this spam.

- Abusing group invites. If your friends are interested, they'll likely join without your "encouragement." And if they don't accept, don't send the group request more than once by asking them to join via email, wall post, or Facebook message.

- Turning your Facebook profile photo into a pitch so that you can gather leads through your Facebook connections. Thanks, but no thanks. Facebook is about real friendships and not about business—at least not primarily.

- Using a fake name as your Facebook name. I can't tell you how many people I've come across with the last name "Com" or "Seo." I'm not adding you unless you can be honest about who you are. Once upon a time, Facebook deleted all of the accounts that portrayed people as business entities or things. I wish Facebook would employ the same tactics again, because I'm not adding a fake identity as a friend.

- Publicizing a private conversation on a wall post. In case it isn't obvious, Facebook wall posts are completely public to all your friends (unless you tweak your privacy settings). Private matters should be handled privately via email or in Facebook private messages.

- Tagging individuals in unflattering pictures that may end up costing your friends their jobs. Visit *http://www.valleywag.com/tech/your-privacy-is-an-illusion/bank-intern-busted-by-facebook-321802.php* to read about a particularly embarrassing incident for one Facebook user who had his picture posted in this manner. Avoid the unnecessary commentary also, especially on your childhood pictures that portray your tagged friends as chubby and not so popular. Further, if your friends request to be untagged, don't make a stink of it.

Figure A-1 shows the number of pending friend requests I have on Facebook.

That said, there's one other rule that some individuals follow, and thus one more sin to add to our Facebook list. I know this isn't the case for all individuals, so your mileage may vary:

- Forgetting that some individuals won't network with you on a "personal" space like Facebook without knowing who you are, even with the proper introduction. If you're looking to establish a professional relationship with someone, consider LinkedIn. Otherwise, consider building up a rapport with an individual before randomly adding her as your friend. Some people require face-to-face meetings before they invite you into their private lives. After all, Facebook was a tool that college students were using before it was open to the public,[2] and some still use it as a purely personal and not a professional tool. LinkedIn is seen as the more professional of the two.

Requests   Ignore All

5 friend suggestions	580 friend requests
5 event invitations	163 group invitations
24 Page suggestions	2 yoville gift requests
1 yoville crew request	1 game invitation
4 facebook game invitations	15 cause invitations
	1 birthday wish invitation
3 petition invitations	11 (lil) green patch requests
1 kidnap! request	1 crazy passover gifts request
1 mafia wars request	2 recruitment requests
1 street racing request	2 birthday requests
1 willow tree angels request	5 relative invitations
	1 add relative request
1 (lil) blue cove request	1 vampires invitation
2 tip'd invitations	1 blood lust invitation
1 world's end request	1 socialmedian invitation
1 willy's sweet shop request	1 good karma from alys request
3 good karma from ross requests	1 drinking request
	1 crusades invitation
1 happy hanukkah! request	2 blog ownership requests
1 google inside invitation	2 birthday requests
1 'my stuff' widget invitation	1 cute pet request
	1 phonebugger invitation
1 yummy gifts request	1 friend's value invitation
1 coolest person invitation	2 sea garden requests
	2 pinky gifts requests
1 send care bears request	2 mob wars invitations
	2 requests
2 rosh hashana cards f requests	1 challah for you request
1 mr men and little mi request	2 butterflies requests
	1 hello kitty request
1 i am gutsy invitation	1 music invitation
1 gary busey's player invitation	2 buy your friends requests

FIGURE A-1. The number of pending requests I have on Facebook

Considering the previous example, I pose a question on Facebook etiquette. Is it appropriate to let these requests sit in pending mode or to reject the friends outright? In many instances, these requests are probably better off sitting indefinitely (and it's healthier than the rejection). Plus, in the future, you may want to respond to that friend request positively.

Twitter

- Following a user and then unfollowing him before he has a chance to follow you. Or unfollowing him as soon as he follows you.

- Mass-following everyone so that you can artificially inflate your numbers as a success metric for influence (and maybe then submitting a press release[3] about it).

- Consistently using your Twitter stream for nothing but self-promotion and ego. Profy highlights this phenomenon quite well at *http://www.profy.com/2008/11/04/quick-tips -for-twitter-spammer-follow-1000-people-give-away-5000-books*.

- Requesting that your friends retweet your tweets on a consistent basis. This is much more bothersome when the request comes via IM or email and not on Twitter itself. The bottom line: *if your content is good enough to stand on its own, it will be retweeted. There is no reason to make a personal request (and if it doesn't stand on its own, it usually doesn't need to be retweeted).*

- Not humanizing your profile. Twitter is also about real relationships. Add an avatar and a bio at the minimum. Let people know who you are. To take it a step further, make it easy for people to contact you outside Twitter if necessary. This is especially important if someone on Twitter needs to reach you but can't direct message you since you're not following her. If she's making the effort, it's probably because she really wants to talk to you. (Was it something you said? Usually.)

- Streaming only your blog's RSS feed on Twitter. If you're following anyone who does this, feel free to take my advice and unfollow him right now. He won't engage with you, so why engage with his narcissistic self-promotion?

- Using Twitter to repeat personal and confidential correspondence. If you're not happy with the way an email communication progressed about a private matter, take it up with the person involved to square things away. Certainly, don't broadcast your dissatisfaction with the turnout to your entire Twitter audience. It looks unprofessional for you and makes you appear untrustworthy.

- Leveraging your Twitter connections to send spam via direct messages to those who follow you. Two days later, you may wonder why they don't follow you anymore.

- Abusing Twitter hashtags during a crisis.[4] It's a shame that the Mumbai attacks happened, but this was not the opportunity to capitalize on your CRM software.

- Using your Twitter feed as a chat room *for conversations that are exclusive in nature* and not as a broadcast medium. It's nice that Twitter lets you use the @ symbol to talk directly to individuals, and that's fine in moderation. As a friend recently said to me, "I'm tired of my Twitter feed being a [*private*] conversation between person X, person Y, and person Z." Why don't the three of you get a room? Twitter user cheapsuits sums it up nicely when he says, "The tweeps that talk everyday to each other about banalities get old." The emphasis here is on chat rooms that exclude other individuals in conversations that do *not* provide value. At all. Ever.

LinkedIn

- Gathering all the email addresses of users you are connected to—even locating email addresses of LinkedIn Group managers—and utilizing this mailing list to promote your own company or service offsite. For example, I manage a few LinkedIn groups, so my email address is far more visible on the site than I'd like. I'm not connected to the LinkedIn individual who spammed me, but he still took the liberty of using my email address for his personal gain in a completely unsolicited fashion. Perhaps this individual lost sight of the fact that LinkedIn is a professional network and not a spam facilitator. Even so, recipients should still be required to opt in.

- Asking for endorsements from individuals you don't know or who didn't do a good job in your employ.

- Writing a recommendation for someone and then firing her just a few days later.[5]

Social News—Digg, Sphinn, Mixx, Reddit, Tip'd (and a Whole Load of Related Sites)

- Submitting only your own articles and posts to social media sites.

- Consistently "taking" (asking for votes) but never giving back. Social news is about reciprocal relationships.[6] Even if the people you are asking for votes of will never actually ask you for votes, a random IM that pops up saying, "Digg this for me," is far more obtrusive than saying "Hey, how's it going?" and having a real conversation first.

- Sharing the same story repeatedly with your friends. Can we say spam? And if you are still being shared with by these members repeatedly, why haven't you unfriended or blocked the offenders on your IM programs or social news sites?

- Submitting a story that is completely off-topic. It's important to understand the communities you contribute to and to understand the rules of the sites that you target.[7] Your story about celebrity cell phones simply does not belong on financial social news site Tip'd, no matter how you try to spin it. And when I, as a moderator, tell you that that the submission is not appropriate for the audience, especially as it has no relevance to the subject matter of the site, don't argue with the decision.

- Using the comments field to drop links, especially to related submissions that were made after the fact.
- On social sites where buries are public (though professional in nature), assuming that it's personal. In a recent instance, a "bury" on a popular social site upset the submitter so much that he resorted to an unprofessional attack on the person who buried the story by blogging about her. Sadly enough, the bury reason (which was public for all to see) was not at all about him but was about the content itself. In social media and in relationships in general, *you should be disagreeing with the statement*. That means that you as the submitter shouldn't assume the burier talking about you as the person who made the statement and implying that the statement is a reflection of a character flaw. The burier didn't like what you said and disagreed. Grow from it. *Don't* turn it into something personal when it clearly *isn't*.

FriendFeed

- Using the service completely for self-promotion. If you're going to claim your social media profile on that totally awesome service, either don't share your feeds at all or interact on a semi-consistent basis. Please? FriendFeed is a service, but it's also a community.
- Cross-posting on all social sites using a site like ping.fm. I don't need to see the same message from you on Twitter, FriendFeed, your Google Talk status, your Facebook feed, and on your dog's scrolling LED collar. Keep the spam broadcasts to a minimum. It's obvious on FriendFeed when this facility is abused.

YouTube

- Asking someone repeatedly to watch your crummy video, subscribe to your channel, and give you a five-star rating.
- Forcing people to subscribe to your YouTube channel by applying an iFrame exploit.[8]

StumbleUpon

- Sending more than one story to your network daily. The key to success is moderation. Excess converts to spam.
- Submitting and reviewing only your own articles. Do you self-promote this often in real life?
- Submitting a story from another social news site to StumbleUpon for more visibility and eyeballs. Once upon a time, I stumbled upon a Digg submission of a Sphinn submission of a blog post. Seriously? Why don't you just submit the blog post directly instead of using

the other sites as conduits? This infraction goes for all social sites that accept submissions, and not just StumbleUpon.

Blogging and Commenting

- Commenting on other articles and using the name "Yellow Brick Plumbing." Isn't your name actually Alan? There's no SEO value to these comments (they're nofollowed by default), and all this approach does is makes you lose credibility in the eyes of the blogger. This isn't the way to *network*!

  > **DEFINITION**
  > A link that has a *nofollow* attribution means that search engines won't actually place link authority when traversing through websites. By default, links are *not* nofollowed, but due to the heavy spam impacting blogs, all comments have this designation. You can turn it off with certain plug-ins, but it's not recommended.

- Using content from another blog without attribution. Sometimes a specific blog will get an exclusive. Then another blog will write on the story using the original blog post as its "source" without attribution. Even popular blogs will rip off stories from lesser-known blogs in their space. Don't let greed get in the way of your own blogging habits, and make sure to link out where appropriate.

- Sending a pitch to a blogger requesting a link exchange even though your site has no relevancy at all to her content. I write about *social media*, people, not about beer bongs. And, well, they say that social media is the new link exchange, so instead of asking for an old-fashioned link (which might have worked in 2002), consider using a more viable strategy for this modern time period.

- Turning a blog into a flame war against someone you don't like. Scott Hendison recounts how forum spam turned into a heated battle that may end up going to the courts, and how the individual responsible for the abuse is not slowing down.[9] If you're wrong, acknowledge the wrongdoing and don't use other blogs to tarnish someone else's image.

Other Social Sites

- Joining a new social network and then inviting everyone you've ever emailed in your lifetime to the service by submitting your entire Gmail address book when the service requests it. Reading the fine print is wonderful—and you should never volunteer your email account's password to the social site anyway. It's also helpful to keep in mind that your email account password should not be the same as your social profiles, and that's not a question of etiquette—it's common sense!

Finally, a Word on Social Media Etiquette in General

You're leaving your digital signature on the Internet right now. Think about the consequences of your engagement on any social site. Racial slurs, criticisms without warrant, and blatant abuse don't work in real life, and they likewise have no place in the social media channels simply because you are far more anonymous on these sites. If you were living in New York and you walked up to a stranger with the same foul-mouthed comments that are rampant on many social media sites, you may never make it home. Consider how your comments will be perceived before you actually post them and think about logic before emotion at all times. Above all, think about maintaining a certain level of professionalism, since people can use whatever you make "permanent" on these sites against you. Not all blogs will remove a comment after you've requested that they do so, simply because you were angry when you wrote it. Before you hit "post," realize that this will be a permanent reflection of your identity and that it may never be erased. It may even be used against you.

Conclusion

Remember that social media communities are real relationships and real conversations, and you should treat them as such. It's not a *me, myself, and I* mentality. It's about the collective, the community, and the common good.

Endnotes

1. *http://www.techipedia.com/2008/social-media-etiquette-handbook*

2. *http://www.techipedia.com/2007/13-reasons-why-i-am-an-obsessive-compulsive-facebook-user*

3. *http://www.sphinn.com/story/89244*

4. *http://www.blogs.zdnet.com/feeds/?p=345*

5. *http://valleywag.gawker.com/5069442/linkedin-recommendation--youre-fired*

6. *http://www.techipedia.com/2007/11-digg-tips*

7. *http://www.techipedia.com/2007/you-cant-own-the-community-without-understanding-them*

8. *http://www.seroundtable.com/archives/018674.html*

9. *http://www.pdxtc.com/wpblog/viruses-and-scams/peak-studios-actually-harming-clients*

Recommended Reading

Recommended Books

Beal, Andy. *Radically Transparent: Monitoring and Managing Reputations Online*, Sybex, 2008

Livingston, Geoff. *Now is Gone: A Primer on New Media for Executives and Entrepreneurs*, Bartleby Press, 2007

Li, Charlene and Bernoff, Josh. *Groundswell: Winning in a World Transformed by Social Technologies*, Harvard Business School Press, 2008

Gillin, Paul. *Secrets of Social Media Marketing: How to Use Online Conversations and Customer Communities to Turbo-charge Your Business*, Quill Driver Books, 2008

Blogroll

Chris Brogan (*http://www.chrisbrogan.com*)

Conversation Agent (*http://www.conversationagent.com*)

Dosh Dosh (*http://www.doshdosh.com*)

Mashable (*http://www.mashable.com*)

Marketing Pilgrim (*http://www.marketingpilgrim.com*)

PR Squared (*http://www.pr-squared.com*)

SEOmoz (*http://www.seomoz.org/blog*)

Social Media Explorer (*http://www.socialmediaexplorer.com*)—Jason Falls

Top Rank Blog (*http://www.toprankblog.com*)

We'd like to hear your suggestions for improving our indexes. Send email to *index@oreilly.com*.

using to find inspirational blog content, 101
Google Blog Search, 101
Google Reader, 90
Google Trends and Google Insights, 102
Graco Baby case study, 66

H

Hacker News, 259
Hashtags (Twitter tool), 146
hi5 social network, 170
Home Depot case study, 70
how-tos, teaching users, 70, 320

I

identity, establishing in social media, 301
iGoogle, 90
IM (Instant Messaging), using in marketing, 322
image-sharing websites
    Flickr, 268–280
    other, 280
images, 268
    (see also photography)
    photoblogs and galleries, 282–283
    using in blog copy, 104
    using to tell a story, 318
ImageShack, 105
Immediate Future, study by, 26
increased traffic (see traffic boost to website)
influence
    blogs as online influencers, 88
    measuring for social media marketing
        campaign, 305
informational social networks (see social networks,
        informational)
interest-based content discovery
    StumbleUpon bookmarking site, 201
    Twine social bookmarking site, 221
Internet evolution, relation to social media
    marketing, 2
iPhone
    third-party applications, 320
    Twitter on, 147
iPod Touch, 295
    Twitter on, 147
Israeli Consulate
    blogs, 39
    use of Twitter as official communications
        channel, 136

J

Jarvis, Jeff, (blogger), 24
JetBlue
    acknowledging mistakes, 23
    use of Twitter for customer service, 132

job-hunting
    using LinkedIn, 163
    using Twitter, 139
journalists using Twitter, 144

K

kirtsy, 254
    popular list on, 238
knowem.com, 309
knowledge exchanges (see question-and-answer
        knowledge exchanges)
knowledge, power of, 193

L

Lifehacker, 60
link building, 3
LinkedIn, 11, 163–169
    Answers, 192
    Applications, 169
    Company Groups, 168
    DirectAds feature, 169
    etiquette for, 331
    Groups feature for user groups, 167
    Questions and Answers section, 164
    Service Providers Recommendations Engine,
        166
links
    driving relevant links to your website, 5
    measuring inbound links for a URL, 304
listening, 39
    (see also conversation, monitoring)
    continual monitoring of community
        conversation, 307
    finding out what community is saying about
        business or competitors, 34
    requirement for, 23
lists
    using in social media marketing, 313
    using on social news sites, 238

M

Mahalo, 180–187
    Answers knowledge exchange, 186
    contributing to, 182
    Greenhouse, for works in progress, 182
    social networking on Mahalo Social, 184
    structure of, 181
marketing tactics, traditional, 322
Mashable.com, 167
measurable goals, 33
MediaWiki, 194
Meetup.com, 311
meetups, organizing on Twitter, 145
Memeorandum, 102

Tamar Weinberg is a writer and Internet marketing consultant specializing in blogger outreach, viral marketing, and social media. She maintains a personal blog about all things social media at *http://www.techipedia.com*. She lives in New York City with her husband and son.

COLOPHON

The cover font is Adobe ITC Garamond. The text font is Linotype Birka; the heading font is Adobe Myriad Condensed; and the code font is LucasFont's TheSansMonoCondensed.

The O'Reilly Advantage

Stay Current and Save Money

Order books online:
www.oreilly.com/store/order

Questions about our
products or your order:
order@oreilly.com

Join our email lists: Sign up
to get topic specific email
announcements or new
books, conferences, special
offers and technology news
elists.oreilly.com

For book content
technical questions:
booktech@oreilly.com

To submit new book
proposals to our editors:
proposals@oreilly.com

Contact us:
O'Reilly Media, Inc.
1005 Gravenstein Highway N.
Sebastopol, CA U.S.A. 95472
707-827-7000 or
800-998-9938
www.oreilly.com

Did you know that if you register
your O'Reilly books, you'll get
automatic notification and upgrade
discounts on new editions?

**And that's not all! Once you've registered
your books you can:**

» Win free books, T-shirts and O'Reilly Gear

» Get special offers available only to registered
O'Reilly customers

» Get free catalogs announcing all our new
titles (US and UK Only)

**Registering is easy! Just go to
www.oreilly.com/go/register**

O'REILLY®

Try the online edition
free for 45 days

"I can think of no one more qualified to bring you kicking and screaming into the 21st century than Tamar Weinberg."
—Dave McClure, 500hats

THE NEW
COMMUNITY
RULES:
MARKETING
ON THE
SOCIAL WEB

O'REILLY®   Tamar Weinberg

Get the information you need when you need it, with Safari Books Online. Safari Books Online contains the complete version of the print book in your hands plus thousands of titles from the best technical publishers, with sample code ready to cut and paste into your applications.

Safari is designed for people in a hurry to get the answers they need so they can get the job done. You can find what you need in the morning, and put it to work in the afternoon. As simple as cut, paste, and program.

To try out Safari and the online edition of the above title FREE for 45 days, go to www.oreilly.com/go/safarienabled and enter the coupon code CYFOHAA.

To see the complete Safari Library visit:
safari.oreilly.com

Safari
Books Online

1801

705C

Shelton State Libraries
Shelton State Community College